SpringerBriefs in Water Science and Technology

More information about this series at http://www.springer.com/series/11214

Noorul Hassan Zardari · Kamal Ahmed
Sharif Moniruzzaman Shirazi
Zulkifli Bin Yusop

Weighting Methods and their Effects on Multi-Criteria Decision Making Model Outcomes in Water Resources Management

Springer

Noorul Hassan Zardari
Sharif Moniruzzaman Shirazi
Institute of Environmental and Water Resources Management (IPASA)
Universiti Teknologi Malaysia
Skudai
Malaysia

Zulkifli Bin Yusop
Water Research Alliance
Universiti Teknologi Malaysia
Skudai
Malaysia

Kamal Ahmed
Department of Hydraulic and Hydrology
Faculty of Civil Engineering
Universiti Teknologi Malaysia
Skudai
Malaysia

ISSN 2194-7244 ISSN 2194-7252 (electronic)
ISBN 978-3-319-12585-5 ISBN 978-3-319-12586-2 (eBook)
DOI 10.1007/978-3-319-12586-2

Library of Congress Control Number: 2014955190

Springer Cham Heidelberg New York Dordrecht London

Printed on acid-free paper

Springer is part of Springer Science+Business Media (www.springer.com)

Contents

About the Authors

Mr. Kamal Ahmed is working as Lecturer at Faculty of Water Resources Management, Lasbela University of Agriculture, Water and Marine Sciences (LUAWMS), Balochistan Pakistan. Currently, he is on leave for completing his Ph.D. at Universiti Teknologi Malaysia. His fields of research include groundwater modeling, climate change and forecasting, and flood warning systems. Mr. Kamal is currently engaged on a research project that investigates groundwater potential in southern Balochistan where he is using GIS approach to located fresh groundwater resources to meet agriculture and drinking demands for the southern Balochistan population.

Dr. Sharif Moniruzzaman Shirazi is working as Associate Professor at Institute of Environmental and Water Resources Management (IPASA), Universiti Teknologi Malaysia. Dr. Shirazi obtained his Ph.D. in the Department of Civil and Environmental Engineering of Saitama University, Japan. He started his career as an Irrigation Engineer. He obtained his B.Sc. and M.Sc. in Irrigation Engineering from Bangladesh Agricultural University. He worked as a JSPS (Japan Society for the Promotion of Science) Postdoctoral Fellow (November 2007–October 2009) at Geosphere Research Institute of Saitama University.

Dr. Shirazi has 17 years research experience in the area of physicochemical properties of soil, agro-ecological aspects of groundwater utilization, groundwater quality for irrigation, and impact of changes on climatic parameters on crop production. He has about 40 technical papers published in national and international journals. He is a member of several institutions including American society of Civil Engineers (ASCE), Engineers Institute of Bangladesh (IEB), Japanese Geotechnical Society (JGS), and Institute of Engineering and Technology (UK).

Prof. Zulkifli Bin Yusop is working as Professor at Universiti Teknologi Malaysia. Currently, he is Dean Water Research Alliance. Prof. Zulkifli has research experience of about 30 years. Prof. Zulkifli has published over 100 articles in top international journals. His fields of research include water resources management, non-point source pollution, surface and groundwater hydrology, catchment management, drinking water quality, deforestation, and flood mitigation and management. Prof. Zul has been invited speaker in various seminars, conference, and workshops. He has supervised more than 20 Ph.D. students in his career. Prof. Zulkifli has secured a large number of national and international grants on different themes of research in water resources and hydrology. He has won many awards in different categories including Best Publication awards.

Dr. Noorul Hassan Zardari works as Senior Lecturer at Institute of Environmental and Water Resources Management (IPASA), Universiti Teknologi Malaysia (UTM), Skudai, Malaysia. His areas of expertise include water resources management, irrigation, groundwater hydrology, decision-making and policy in water resources allocations, integration of land and water resources in watersheds, and urban water supply and management. He is author of over 40 publications.

Dr. Noorul has organized many publication workshops in which academicians, researchers, and postgraduate students from different faculties of the Universiti Teknologi Malaysia were trained on enhancing writing and publication skills. The major objectives of such workshops were to promote publication culture in the university and improve quality of the research. Dr. Noorul has reviewed about 50 journal articles which were submitted into indexed, non-indexed journals and conferences. He is also serving as Chairman of the Publication Committee, Water Research Alliance, Universiti Teknologi Malaysia where many students are assisted in their journal articles and theses write up.

Abstract

This book presents survey results on priority ranking of watershed management criteria. This survey was completed by 30 undergraduate and postgraduate students from Faculty of Civil Engineering, Universiti Teknologi Malaysia. The students were registered in various programs run by the Faculty in Semester 1, 2012. The applications of three weighting methods namely rating, ranking, and ratio are discussed in the book. We have also done data mining on some keywords using three popular scholar databases. These databases include *sciencedirect*, *scopus*, and *sciverse*. Four abbreviated keywords (MCDM, MCDA, MCA, MADM) used to represent multi-criteria decision-making were used and these three databases were searched for different popular weighting methods for a period of 13 years (2000–2012). The findings of data mining are presented in this book. Overall, this book presents a review of weighting methods applied in various multi-criteria decision-making (MCDM) methods.

Keywords Weighting methods · MCDM · Multi-criteria decision-making · Ranking · Water resources

Chapter 1
Introduction

1.1 Introduction

For last more than 20 years, multi-criteria decision making (MCDM) models have been applied to the field of natural resources management. In literature, eight areas of application of MCDM models are identified in water resources management: catchment management; ground water management; infrastructure selection; project appraisal; water allocation; water policy and the planning of supply; water quality management; and marine protected area management. MCDM models can provide solutions for complex water decision-making problems. In majority MCDM models, assigning weights to the evaluation criteria is an important step. For that, various weighting methods have been proposed in literature and applied for solving different MCDM problems. These weighting methods are classified in different ways: algebraic or statistical, decomposed or holistic, direct or indirect, and compensatory or non-compensatory. Compensatory weighting methods are used in Multi Attribute Utility (MAU) methods and non-compensatory weighting methods are used mainly in outranking methods (e.g. ELECTRE and PROMETHEE). Point allocation, ranking methods, rating methods, pairwise comparison, and trade-off analysis are some popular weighting methods. Each weighting method, however, differs in terms of accuracy, ease of use, complexity for users, and theoretical foundations and produce different sets of criteria weights. The decision for selection of a proper weighting method is a crucial step in solving a multi-criteria decision problem. Many researchers have dismissed the difficulty in measuring and meaning the criteria weights and assumes that the meaning of criteria importance is well understood by all decision makers. However, true meaning and the validity of criteria weights obtained by using different weighting procedures are important for avoiding any misuse of the MCDM models and getting reliable model results.

The main objective of the project is to evaluate different weighting methods based on their subjective and objective inputs and their influence on the model results. For that, we have hypothesized a watershed for which management plan is

N.H. Zardari et al., *Weighting Methods and their Effects on Multi-Criteria Decision Making Model Outcomes in Water Resources Management*,
SpringerBriefs in Water Science and Technology, DOI 10.1007/978-3-319-12586-2_1

need to be developed. A priority ranking of environmental and socio-economic criteria are intend to be developed from the survey participants' preferences on various watershed management criteria and weights for criteria are elicited for further usage in MCDM models. In this study, a questionnaire was designed and administered to the undergraduate and postgraduate students of Faculty of Civil Engineering, Universiti Teknologi Malaysia. In the survey, the participants were asked to answer the question "Rank the decision criteria according to their importance (most important to least important) for a watershed management problem". The participants were asked to show their preferences by using three different weighting methods i.e. ranking, rating, and ratio weighting methods.

1.2 Problem Background

Multi-criteria decision making (MCDM) models have been applied to decisions on the management of natural resources over the last 20 years or so (Hajkowicz and Higgins 2008). As for water resources management, Hajkowicz and Collins (2007) identified eight areas of application: catchment management; ground water management; infrastructure selection; project appraisal; water allocation; water policy and the planning of supply; water quality management; and marine protected area management. Multi-criteria decision making models provide solutions for complex water decision-making problems (Goncalves and Pereira 2009; Silva et al. 2010). MCDM models are used to evaluate a finite set of alternatives with respect to multiple criteria. Alternatives are first evaluated with respect to each of the criteria to obtain criterion priority scores which are then aggregated into overall preference values (Choo et al. 1999). These scores and overall preference values may be in ordinal, interval or ratio scales. Many different methods have been proposed for assessing criteria weights which are then used to aggregate criterion priority scores. Thus, the true meaning and the validity of these criteria weights are important for avoiding misuse of the MCDM models. Unfortunately, criteria weights are often misunderstood and misused (Zhang and Wang 1992).

There are various methods to determine criteria weights, which can be classified in different ways: algebraic or statistical, decomposed or holistic, direct or indirect, and compensatory or non-compensatory. Direct methods require the respondents to compare two criteria in terms of ratio judgments and indirect procedures ask them to express preference judgments to derive criteria weights. Compensatory weighting techniques are used in Multi Attribute Utility (MAU) methods, while non-compensatory ones are used mainly in outranking methods (e.g. ELECTRE and PROMETHEE). Popular weighting methods include point allocation, ranking methods, rating methods, pairwise comparison, and trade-off analysis. Each weighting method differs in terms of accuracy, ease of use, complexity for users, and theoretical foundations and produce different sets of criteria weights. Weighting methods are either oversimplified, lacking any theoretical foundation (Hokkanen and Salminen 1997). The decision for selection of a proper weighting method is important in solving a particular multi-criteria decision making problem.

Many researchers, however, have dismissed the difficulty in measuring and meaning the criteria weights and assumes that the meaning of criteria importance is transparent and well understood by all decision makers (Choo et al. 1999).

In this study, we are providing a thorough discussion on advantages and disadvantages of some popular weighting methods. Issues like subjectivity and misuse of MCDM models in previous studies are also discussed in detail. Different non-compensatory weighting methods (i.e. direct point allocation, spontaneous ranking, and ratio methods) for eliciting criteria weights from decision makers and stakeholders are applied in a survey questionnaire in which a hypothetical watershed management problem was presented to the survey participants. The participants of the survey were asked to answer the question "Rank the decision criteria that are assumed to be important for managing any watershed". The participants arranged the watershed management criteria in order of their priority (most important to least important). After they listed them, they assigned weights to all criteria starting with a weight of 100 % for the most important criterion (Rank 1). Based on different effects of weighting methods on final results of a MCDM model and influence of subjectivity on model outcomes can be evaluated in future studies and a decision framework can be developed to assist decision makers in selecting most suitable weighting method for solving a watershed management problem.

1.3 Problem Statement

There are various methods to determine criteria weights, which can be classified in different ways: algebraic or statistical, decomposed or holistic, direct or indirect, and compensatory or non-compensatory. Direct methods require the respondents to compare two criteria in terms of ratio judgments and indirect procedures ask them to express preference judgments to derive criteria weights. Compensatory weighting techniques are used in Multi Attribute Utility (MAU) methods, while non-compensatory ones are used mainly in outranking methods (e.g. ELECTRE and PROMETHEE). Popular weighting methods include point allocation, ranking methods, rating methods, pairwise comparison, and trade-off analysis. Each weighting method differs in terms of accuracy, ease of use, complexity for users, and theoretical foundations and produce different sets of criteria weights. Weighting methods are either oversimplified, lacking any theoretical foundation (Hokkanen and Salminen 1997). The decision for selection of a proper weighting method is important in solving a particular multi-criteria decision making problem. Many researchers, however, have dismissed the difficulty in measuring and meaning the criteria weights and assumes that the meaning of criteria importance is transparent and well understood by all decision makers (Choo et al. 1999).

In this study, we have summarized all above issues related to the weighting methods and have illustrated how the selection of weighting method for a particular multi-criteria problem is important and how the final results of multi-criteria decision models were dependent on the use of different weighting method.

1.4 Objectives of the Study

The main objective of the study was to critically examine the properties of some popular weighting methods which are being currently used to establish relative importance of criteria/attributes in multi-criteria decision making (MCDM) methods to aid decision-makers in solving real-world problems. The specific objectives of the study are as follows:

- To investigate how different weighting methods affects the quality of decisions based on multi-criteria decision making models
- To develop a priority ranking of weighting methods based on their easiness, accuracy, and strong theoretical foundation
- To investigate how (and to what degree) the subjective weights affect the outcomes of the multi-criteria decision making models.

1.5 Scope of the Study

The study focuses on the weighting methods which have been frequently used to know the importance of multiple criteria in multi-criteria decision making methods. In this study, our more focus was on those weighting methods which were used in solving multi-criteria decision problems related to water resources and hydrology. The data was extracted from three popular online databases (sciencedirect, sciverse, and scopus) for a period of only 13 years (i.e. 2000–2012). A survey on three weighting methods (rating, ranking, and point allocation) was completed from small groups of undergraduate and postgraduate students registered at Faculty of Civil Engineering, Universiti Teknologi Malaysia, Skudai campus.

1.6 Significance of the Study

Literature on weighting methods is very rich. However, summarizing them according to their effects on final model outputs is missing in the literature. We have attempted to quantify these effects through this study. The findings of the study could assist decision makers to select the best weighting method for their studies according to human resources, funds, and time availability factors. We emphasize that the weighting methods applied in solving multi-criteria problems in water resources and hydrology were only listed in this study. Weighting methods used in solving problems outside of water resources and hydrology are excluded from this study.

1.7 Chapter Summary

This chapter presents an overview of this study. It includes a brief introduction, background of the study, the problem statement, the study objectives, the scope of the study and its significance.

1.8 Report Organization

This report is organized into four chapters. This chapter presents an introduction and the background of the study, which is followed by the problem statement, scope of the study and the significance of the study. Chapter 2 is limited to literature review on weighting methods applied in multi-criteria decision making methods. A brief introduction of multi-criteria decision making methods is also provided in this chapter. Chapter 3 mostly discusses the methodological part of the study. Here we have given the procedure for extracting data using three main online databases. A framework of the applied methodology is also presented in this chapter. Results of data analysis and interpretation of data mining and survey results are also given in this chapter. Chapter 4 covers conclusions and main findings of the study. Recommendations for future work are also given in this chapter.

Chapter 2
Literature Review

2.1 Introduction

This chapter is divided into two main parts. The first part of the chapter consists of literature review on different types of multi-criteria decision making (MCDM) methods. Literature review on the application of MCDM methods in different fields is also provided in the first part of the chapter. The second part of the chapter presents literature review on weighting methods and different types of weighting methods. We have also summarized advantages and disadvantages of various weighting methods in this part of the chapter. The criteria used to select popular weighting method for a particular water resource or hydrology study are also discussed in the second part of the chapter. A brief review report on various applications of the weighting methods in different MCDM methods is also given in this part of the chapter.

2.2 Decision-Making Process

In most of the cases, decision-making process takes the steps shown in Fig. 2.1. In the first step, problem in hand is clearly defined. Some other important requirements are then listed on which the solution of multi-criteria model was dependent. In the third step, objectives or goals of the multi-criteria problem are established. Fourth step of the decision-making process deals with the establishment of alternatives which are going to be considered in a decision-making process with objective to choose the best alternative. In Step 5 of the decision process, evaluation criteria are decided. The criteria should satisfy some previously fixed standards. For example, the chosen criterion may change its value in space and time. The sixth step of the process is very important as it involves the selection of an appropriate multi-criteria decision making method for solving the problem in hand. Later the chosen MCDM

N.H. Zardari et al., *Weighting Methods and their Effects on Multi-Criteria Decision Making Model Outcomes in Water Resources Management*,
SpringerBriefs in Water Science and Technology, DOI 10.1007/978-3-319-12586-2_2

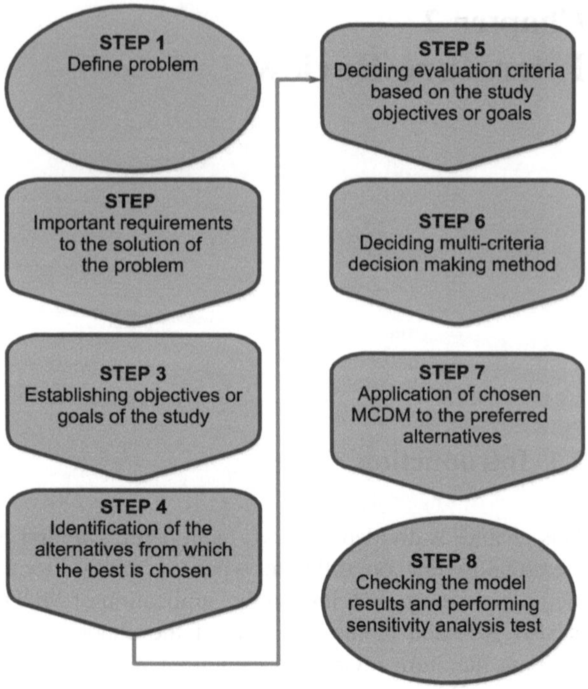

Fig. 2.1 Decision making process

method is applied to the list of alternatives which was finalized in Step 4 of the decision process. Final step of the decision-making process is checking the results of the model and performing sensitivity analysis test.

It is important to say that the decision-making process normally flows from top to bottom, but it may return to any of the previous steps if new information was later found.

Yoe (2002) describes the multi-criteria decision making process as:

1. Define multi-criteria problem and objectives explicitly.
2. List and describe alternatives for meeting objectives or goals.
3. Define criteria/attributes/performance indicators to measure performance of alternatives.
4. Carry out studies to gather data and evaluate criteria.
5. Prepare a decision matrix by arranging alternatives against criteria.
6. Elicit criteria subjective or objective weights for criteria.
7. Rank alternatives and communicate results with interest groups.
8. Decision-makers make decisions with input of interest group and get MCDM results.

These steps are shown in Fig. 2.2.

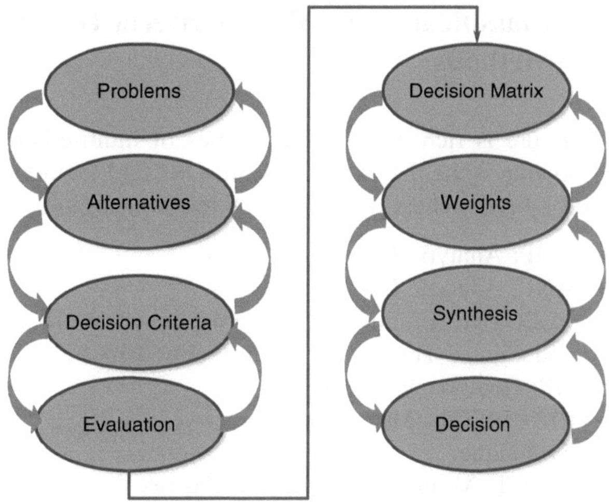

Fig. 2.2 The iterative steps of MCDM (after Yoe 2002)

2.3 Multi-Criteria Decision-Making

International Society on Multiple Criteria Decision Making defines MCDM as "The study of methods and procedures by which multiple and conflicting criteria can be incorporated into the decision process." The development of multi-criteria decision-making began in 1971. The main objective of MCDM is to provide decision-makers with a tool in order to enable them to advance in solving a multi-criteria decision problem, where several conflicting criteria are taken into account.

Roy (1996) defines a multi-criteria decision problem as being a situation in which, having defined a set A of actions and a family F of criteria, the decision maker wishes: to determine a subset of actions considered to be the best with respect to F (choice problem); to divide A into subsets according to some norms (sorting problem); to rank the actions of A from the best to worst (ranking problem); to describe actions and their consequences in a formalized and systematic manner, so that decision-makers can evaluate those actions (description of issue) (Schramm and Morais 2012).

In literature, many terms have been used for MCDM and these terms are given as below:

- Multi-Criteria Decision Analysis (MCDA)
- Multi-Objective Decision Making (MODM)
- Multi-Attributes Decision Making (MADM)
- Multi-Dimensions Decision-Making (MDDM)

2.4 Classification of Multi-Criteria Decision-Making Methods

Literature is rich with different types of multi-criteria decision-making methods. Following is the list of some popular MCDM methods which have been frequently used by researchers to solve some real-world multiple criteria problems:

- AHP: Analytic Hierarchy Process
- ANP: Analytic Network Process
- ELECTRE: Elimination Et Choix Traduisant la Realite (French)—(Elimination and Choice Translating Reality) (English)
- GP: Goal Programming
- MACBETH: Measuring Attractiveness by a Categorical Based Evaluation Technique
- MAUT: Multi-Attribute Utility Theory
- MAVT: Multi-Attribute Value Theory
- PROMETHEE: Preference Ranking Organization Method for Enrichment Evaluation
- TOPSIS: Technique for Order Preference by Similarity to Ideal Solution
- WSM: Weighted Sum Model

The specialists have divided multi-criteria decision-making methods into three categories, whose purpose is to bring the MCDM methods together according to some similarities, namely: (i) multiple attribute theory; (ii) outranking methods; (iii) interactive methods. Roy (1996) classifies them as follows: (i) unique synthesis criterion approach, eliminating any incomparability; (ii) outranking synthesis approach, accepting incomparability; (iii) interactive local judgment approach, with trial-error interaction (Schramm and Morais 2012).

- **Unique synthesis criterion approach**: It consists of aggregating the different points-of-view into a unique function which will be optimized. For example, MAUT (Multi-Attribute Utility Theory; Keeney and Raiffa 1976), SMART (Simple Multi-Attribute Rating Technique) family (Edwards 1977; Edwards and Barron 1994) and AHP (Analytic Hierarchy Process) (Saaty 1987).
- **Outranking synthesis approach**: It consists in the development of a relationship called an outranking relationship, which represents the decision-maker's preferences, the relationship being explored in order to help the decision-maker solve his/her problems. Examples: ELECTRE (Elimination and Choice Translating Algorithm) (Belton and Stewart 2002; Roy 1996; Vincke 1992) and PROMETHEE (Preference Ranking Organization Method for Enrichment Evaluation) (Brans and Vincke 1985).
- **Interactive local judgment approach**: This proposes methods which alternate calculation steps, giving successive compromising solutions, and dialog steps, leading to an extra source of information on the decision-maker's preferences (Vincke 1992).

Classification of MCDM methods is shown in Fig. 2.3.

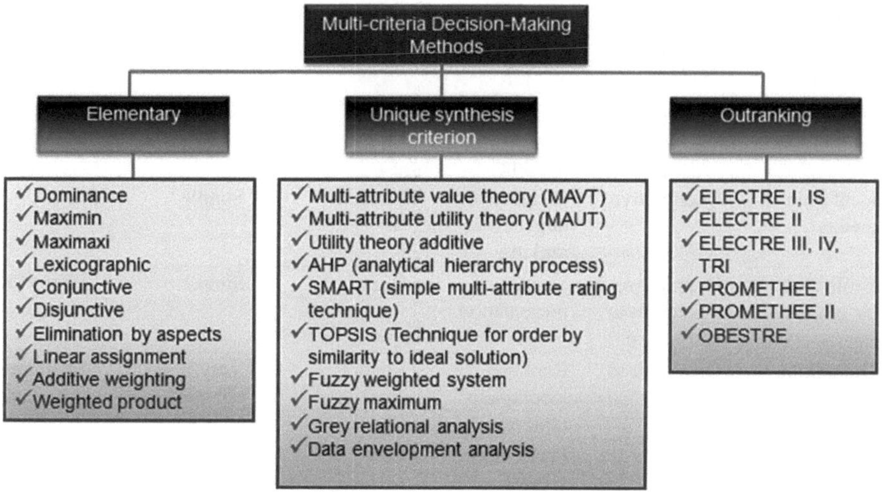

Fig. 2.3 Classification of MCDM methods

2.5 Characteristics of Different Multi-Criteria Methods

Not all MCDM methods are recommended for solving any multi-criteria decision problem. Some MCDM methods can only take quantitative data to process with evaluation phase of the decision-making and some can work with both types of data (quantitative and qualitative). There are also some other characteristics of multi-criteria decision-making methods, e.g. transparency and cost (Table 2.1).

The type of available information will largely determine which MCDM method could be used for a particular multi-criteria problem. Most quantitative methods produce performance scores as well as a ranking. In addition to a ranking, weighted performance scores provide information on the relative performance of the alternatives. Comprehensiveness is achieved if all the information is presented to decision-makers, while presenting a final ranking, or even only one best alternative, results in maximum simplicity and possibly an oversimplification.

Graphic or other presentations of the information take an intermediate position. Although a complete ranking provides maximum simplicity, in aggregating all information into a final ranking, priorities need to be included and a decision rule needs to be selected (RPA 2004).

Transparency is low across a number of the methods, suggesting that such methods should not be used if many stakeholders are involved in or concerned with decision-making. Computation is complex in some of the methods. Since software is generally available to support the use of the methods, this is in itself not an

Table 2.1 Characteristics of different multi-criteria methods (after RPA 2004)

MCDM method	Information	Result	Transparency	Computation	Costs
Weighted summation	Quantitative	Performance scores/ ranking	High	Simple	Low
Ideal point method	Quantitative	Distance to target/ ranking	Medium	Simple	Low
Evaluation by graphics	Qualitative, quantitative and mixed	Visual presentation	High	Simple	Low
Outranking methods	Quantitative	Ranking/ incomplete ranking	Low	Very complex	Medium
Analytical hierarchy process (AHP)	Qualitative	Performance scores/ ranking	Low	Complex	Medium
Regime method	Qualitative, quantitative and mixed	Ranking/ probability	Low	Very complex	Low
Permutation method	Qualitative	Ranking	Low	Very complex	Medium
Evamix method	Mixed	Ranking	Low	Simple	Low

important issue. The costs of adopting methods based on the use of value/utility functions are likely to be higher than those associated with the use of AHP and outranking methods. These additional costs result from the involvement of an expert in the assessment procedure (RPA 2004).

2.6 Strengths and Weaknesses of MCDM Methods

Multi-criteria decision-making methods have been criticized by many researchers for their property of being prone to manipulation which may lead to a false sense of accuracy. One the other hand, supporters of the MCDM approach claim that MCDM provides a systematic, transparent approach that enhances objectivity and generates results which can be trusted with a reasonable satisfaction (Janssen 2001; Macharis et al. 2004). The main elements of criticism to MCDM approach were summarized by Mutikanga (2012) and given as follows:

1. **Aggregation algorithms**: Different MCDM methods yield different outcomes when applied to the same multi-criteria problem. The selection of an appropriate MCDM method from a long list of MCDM methods is often not straight forward and may possibly control the final outcome of the decision-making process.
2. **Compensatory methods**: Complete aggregation methods of the additive type (e.g. AHP) allow for trade-offs between good performance on one criterion and poor performance on some other criterion. Often important information is lost by such aggregation (e.g. in PROMETHEE II complete ranking). For example, poor performance on water quality could be compensated with good performance on investment cost. The underlying value judgments of the aggregation procedure are therefore debatable and probably not acceptable from the public health and regulatory point of view. A multi-criteria problem is mathematically ill-defined since an action a may be better than an action b according to one criterion and worse according to another. This is because complete axiomatization of multi-criteria decision theory is very difficult (Munda et al. 1994).
3. **Elicitation process**: The way subjective information (weights and preference thresholds) is elicited is not trivial and is likely to influence the results.
4. **Incomparable options**: As the purpose of all MCDM is to reduce the number of incomparability, MCDM problems are often reduced to single-criterion problems for which an optimal solution exists completely changing the structure of the decision problem which is not realistic. In addition alternatives are often reduced to a single abstract value during data aggregation resulting in loss of useful information. To a lay person it may be easy to understand the cost of an alternative in monetary values rather than an abstract value indicating that option A is better or worse than option B by a value of say 0.45.
5. **Scaling effects**: Some MCDM methods derive conclusions based on scales in which evaluations are expressed which is unacceptable. For example if two strategy options (A and B) with the same weight (0.5) have different costs (A = 10,000, B = 18,000) and their impact on water quality improvement is (A = 0.2, B = 0.8), their overall performance would be (A = 5000.1 and B = 9000.4). If costing were scaled back to a 0–1 scale, then the relative importance of the two criteria would be better represented.
6. **Problem structuring**: Results could be manipulated by omission or addition of some relevant criteria or options. MCDM methods have been reported to suffer from rank reversals by introduction of new options (De Keyser and Peeters 1996; Dyer 1990).
7. **Additional required information**: Depending on how much additional information is required by the different MCDM methods, "black box" effects are likely to occur thus compromising the ability of the decision-maker to clearly follow the decision process and evaluate the results.
8. **Uncertainty**: The results are often provided to two decimal places which give a false sense of accuracy considering the uncertainties in the input data used and their error propagation in the model. Uncertainty is also inherent in the decision-making process in that it is difficult to quantify and represent performance of most options by a single value.

RPA (2004) divides multi-criteria decision-making methods into different categories and gives brief description of each MCDM method and discusses some key issues associated to some MCDM methods (see Table 2.2).

Based on different types of information (e.g. information on criteria, information alternative), Hwang and Yoon (1981) have categorized different MCDM methods (see Fig. 2.4).

Hajkowicz et al. (2000) present classification of input data (i.e. quantitative, qualitative, and mixed) (Fig. 2.5).

2.7 How to Select an Appropriate MCDM Method

Abrishamchi et al. (2005) state that selecting an appropriate MCDM from a long list of available MCDM methods is a multi-criteria problem itself. There is no single MCDM method which can be superior method for all decision-making problems. Different researchers have different views on this issue. Guitouni and Martel (1998) argue that different MCDM methods will yield different recommendations while Hajkowicz and Higgins (2008) argue that the ranking of decision alternatives is unlikely to change noticeably by using a different MCDM method provided ordinal and cardinal data are handled correctly. However, Guitouni and Martel (1998) have developed some guidelines which can still be helpful in selecting an appropriate MCDM method. A recent review of MCDM for water resource planning and management has shown that MCDM is mostly used for water policy evaluation, strategic planning and infrastructure selection (Hajkowicz and Collins 2007).

2.8 The Role of Weights and Their Interpretation in MCDM Methods

Many MCDM methods (e.g. ELECTRE I, II; PROMETHEE) use criteria weights in their aggregation process. These weights to criteria play an important role for measuring overall preferences of alternatives. Because of having different aggregation rules, MCDM methods use these weights in different ways. For that, different weighting methods have been developed to use them in different MCDM methods. It is very importance that the decision-maker (DM) understands the true meaning of these weights. Choo et al. (1999) suggested that the questions posed to decision-maker in the weight elicitation process must convey the correct meaning of criteria weights. The questions posed to the decision-maker should be direct and simple but not compromising the underlying theoretical validity (Choo et al. 1999). In MCDM

Table 2.2 Brief summary of main categories of multi-criteria decision-making methods and key issues arising from their application (after RPA 2004)

Method	Description	Key issues
Simple methods		
Pairwise comparison	Involves listing the criteria and comparing alternatives in pairs against each of these criteria, indicating a preference for one alternative over another. The results are recorded in a table. An overall preference is then identified. The decision makers must make a judgment on the relative importance to be assigned to the different criteria and thus to determine the "best" alternative	No attempt is made to incorporate the relative importance of different criteria
		Undertaking the comparisons and ensuring consistency becomes increasingly complex as the numbers of criteria and alternatives increases. Applying pairwise comparison technique in such cases can only effectively be achieved through the use of the more sophisticated mathematical approaches (such as the analytical hierarchy process)
Ranking	Involves the ordering of alternatives into ranks using verbal, alphabetical or numerical scales and provides an indication of relative performance. Expert opinions are used to decide the order of preference for different alternatives	When used on their own, they provide little information on the degree or magnitude of any differences in impact between alternatives. These methods hide any uncertainty that may exist as to the extent of such differences. In addition, when there are several alternatives under consideration, it may be difficult to select a preferred alternative. Also, there is a tendency to add ranks together, a mathematical operation which is invalid unless it is assumed that decision-makers place an equal value on impacts falling under the various criteria and that all trend scores or ranks reflect proportional changes in level of impact (i.e. +++ is three times better than +)
	These methods provide a simple means of evaluating the performance of different alternatives over a range of different criteria	

(continued)

Table 2.2 (continued)

Method	Description	Key issues
More complex methods		
Multi-attribute utility analysis	All methods evaluate the attractiveness of alternatives in terms of three discrete elements: The consequences of the alternatives in terms of the decision criteria The relative preference the criteria are denoted in terms of weights; and If the effect scores are measured on different scales, they must be standardized to a common dimensionless number	The computational requirements of the methods are fairly simple. The difficulty with weighted summation lies in choosing a good standardization method and attribution of the weights. In general, linear standardization can be seen as an approximation of more complex non-linear value functions. If the range of scores is not too large for all criteria, linear standardization will be sufficiently accurate. If the range of scores is large or very sensitive, expert judgment should be used to determine the shape of the value function. However, this is time consuming
	The methods vary with respect to their treatment of scores and weights	
Weighted summation	Weighted summation is the simplest form of multi-attribute utility analysis that applies a linear relationship. It involves standardizing the scores across all criteria, assigning preference weights, multiplying the weights by the scores, adding up the resulting scores to obtain total weighted scores for each alternative, and determining the ranking or the total weighted scores	It is less suitable for processing qualitative information. In practice, this disadvantage is not very significant. With a well-chosen method of standardization, this underlying quantitative scale can be used in the weighted summation of these scores
	Although the method requires quantitative information on scores and priorities, only the relative values are used in the assessment. The method does, however, provide a complete ranking of alternatives and information on the relative differences between alternatives	The results of weighted summation can easily be presented in bar graphs. A relative contribution of all criteria or objectives to the overall rankings of the alternatives. These rankings can be used to analyze the sensitivities of ranking of alternatives to overall uncertainty in both effects and priorities
Ideal point method	Based on the concept of value or utility maximization, the Ideal Point Method ranks the alternatives in terms of the degree to which they achieve a pre-specified target or ideal situation (i.e. their distance from the target outcome). Alternatives that are closer to the ideal are preferred to those that are further away	
	By using a scaling coefficient it allows for the inclusion of the relationship between relative size of the effect and weight into the decision rule	

(continued)

Table 2.2 (continued)

Method	Description	Key issues
	The ideal point method provides a complete ranking of alternatives and information on the relative distance of each from the ideal solution	
Evaluation by graphics	Computerized models often provide a graphical interface to facilitate the development and analysis of a decision problem using multi-criteria analysis. One of the key benefits of using a graphical output is that it enables the analyst or decision maker to see easily the relative performance of alternatives under different weighting systems	
Outranking methods	The various ELECTRE methods are the most important representatives of the class of outranking methods. These methods are widely used. They first translate criterion scores to an outranking relationship and then analyze this relationship	The analysis using these methods need to be trained well in their use since interaction with the decision-makers is a complicated task. The results are very sensitive to the level of the thresholds used to define the concordance and discordance relationships
	The variant described here is known as ELECTRE II. It is based on a pairwise comparison of the alternatives, using only the interval character of the scores in the evaluation of the effects table. The basic idea is to measure the degree to which scores and their associated weights confirm or contradict the dominant pairwise relationships among alternatives. Within this method, a dominance relationship for each pair of alternatives is derived using two indices, one index indicating concordance and the second index indicating discordance. The concordance index represents the degree to which alternative i is better than alternative j. The discordance set D_{ij} is defined as the set of evaluation criteria for which alternative i is worse than alternative j. It reflects the idea that, beyond a certain level, bad performance on one criterion cannot be compensated for by good performance on the other criterion. Thresholds supplied by the decision maker	In addition, the procedure used to generate the final ranking does not always result in a complete ranking of alternatives. In some cases one or more alternatives cannot be ranked or two partial rankings are produced. As a result, it is likely that the complexity of the method makes it less transparent

(continued)

Table 2.2 (continued)

Method	Description	Key issues
Qualitative methods		
Analytical hierarchy process (AHP)	Pairwise comparisons provide the basis for the AHP. AHP structures the decision problem into levels that correspond to a decision maker's understanding of the situation: goals, criteria, sub-criteria, and alternatives, so that the decision maker can focus on smaller sets of decisions	AHP is widely used all over the world. Many applications can be found in the literature and lively discussions on its theoretical validity can also be found. Although some controversy surrounds the theoretical basis of the method, it is easy to use and produces results that match the intuitive expectations of the users. Despite its ease of use, the procedure for processing information obtained from the decision maker is far from transparent. This makes the method less suitable for situations with many stakeholders. In addition, for AHP, the number of pairwise comparisons to be made increases rapidly with the number of criteria. Therefore the use of a hierarchal structure of goals, sub-goals and criteria, may be a better option
	The aim of this method is to derive quantitative scores and weights from qualitative statements on the relative performance of alternatives and the relative importance of criteria obtained from comparison of all pairs of alternatives and criteria	
	If the judgments supplied by the decision maker are completely consistent, one row of the comparison matrix A would be enough to produce all relative weights	
	It should be noted, that the AHP method can be used not only to assess relative criteria weights but also to assess the performance of alternatives through pairwise comparisons. The resulting tables of pairwise comparisons are translated to weights and scores using the Eigenvalues of these tables	Decision situations where only strictly ordinal information is available are rare. In these rare cases the Regime method and Permutation method can be used to process this information
Regime method	The Regime method is also based on pairwise comparisons of alternatives. For each criteria all pairs of alternative receives +1, the worst −1 and both alternative receive 0 if they are the same. These scores are then combined with quantitative information on weights attached to the criteria to determine which of the two alternatives is preferred if all criteria are taken into account simultaneously. This is straightforward if quantitative weights are available	

(continued)

Table 2.2 (continued)

Method	Description	Key issues
Permutation method	The Permutation method addresses the following question: which, of all possible rank orders of alternatives, is most in harmony with the ordinal information contained in an effects table?	
	In the case of l alternative, the total number of possible permutations is equal to l!. Each permutation can be numbered as p(p = 1, l!). Each rank order from the permutations is then confronted with the ordinal information contained in each of the rows of the effects table. Rank correlation coefficients are then used to compute the statistical correlation between the l! rank orders and the j columns of the effects table. This results in a large number of rank correlation coefficients. The weighted sums of the rank correlation coefficients are used to determine the most attractive of the l! permutations	
The Evamix method	The Evamix method is designed to deal with an effects table containing both qualitative and quantitative criteria. The set of criteria in the effects table is divided into a set of ordinal criteria O and as set of quantitative criteria Q. For both sets, dominance criteria are calculated	
	The method requires quantitative weights but can be used in combination with any of the methods dealing with ordinal priority information. A total dominance score is found by combining the indices α_{ij} and β_{ij} calculated separately for the qualitative and quantitative scores. To be able to combine α_{ij} and β_{ij} both indices need to be standardized. The most straightforward standardization divides qualitative indices by the absolute value of their sum and does the same with quantitative indices. The total dominance score is calculated as the weighted sum of the qualitative and quantitative dominance scores	

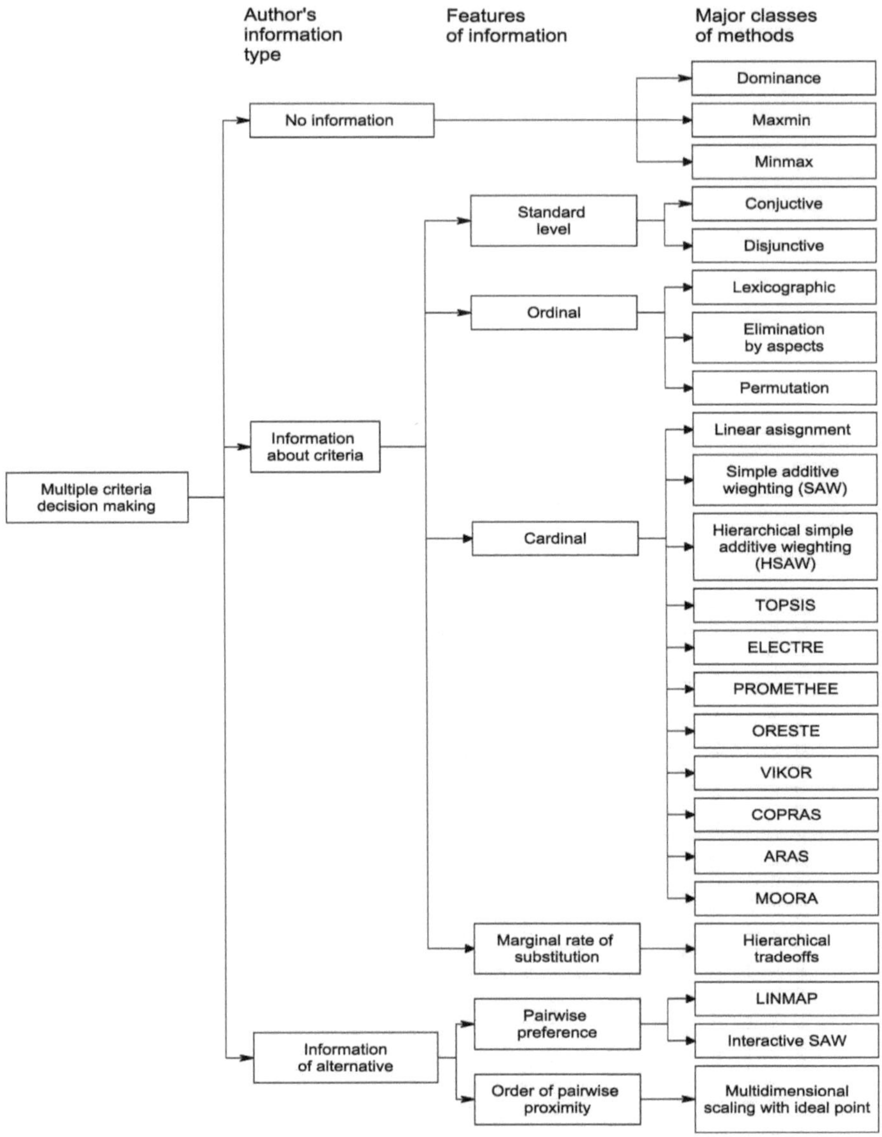

Fig. 2.4 Grouping of multiple criteria decision making methods (after Hwang and Yoon 1981)

literature, the criteria weights $w_1, \ldots, w_p$ have been given a diverse array of plausible interpretations associated with the following (Choo et al. 1999):

1. marginal contribution per unit of $z_k(x)$ or $r_k(x)$,
2. indifference trade-offs or rates of substitution,
3. gradient of the overall value function $U(Z(x))$ or $U(R(x))$,

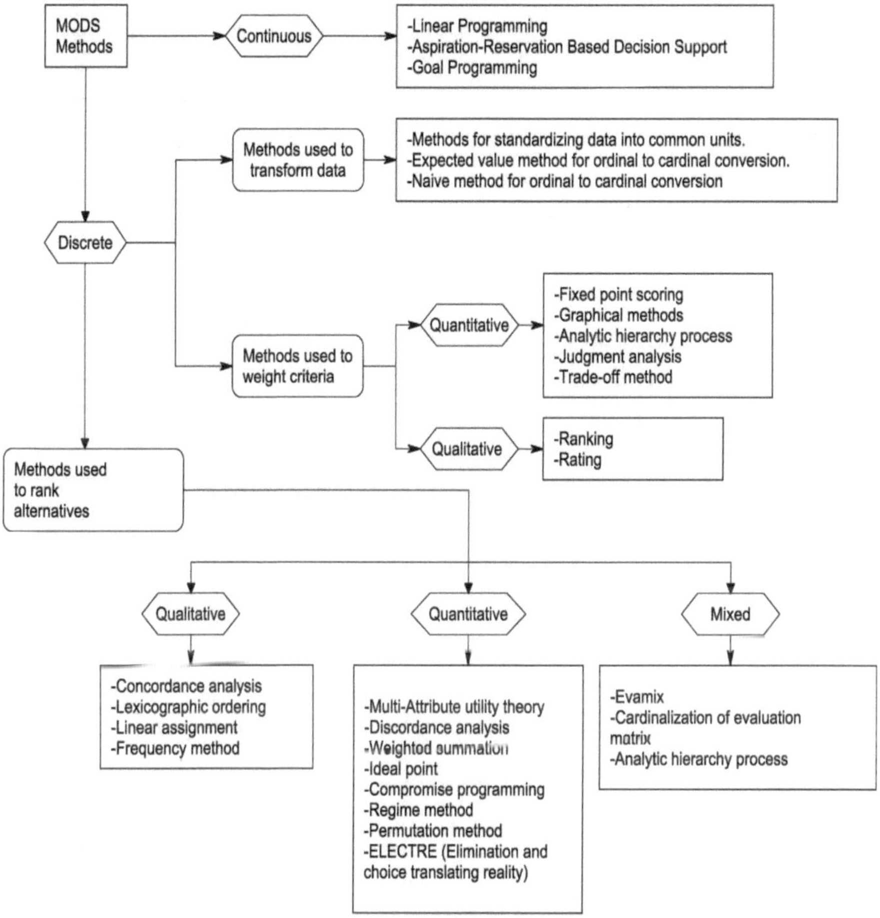

Fig. 2.5 Classification of MCDM methods (after Hajkowicz et al. 2000)

4. scaling factors converting into commensurate overall value,
5. $U(Z(x))\,\theta = (\Sigma\, w_k z_k(x))\,\theta$ or $U(R(x))\,\theta = \Sigma\, w_k r_k(x)\,\theta$ is linear,
6. relative contribution of the average criterion specific scores,
7. discriminating power of the criteria on the alternatives,
8. relative contribution of swing from worse to best on each criterion,
9. vote values in binary choices,
10. relative contribution of the criteria at the optimal alternative,
11. parameters used in interactive optimization,
12. relative information content carried in the criteria,
13. relative functional importance of the criteria.

Table 2.3 presents summary of the interpretations of criteria weights in MCDM methods.

Table 2.3 Meaning of criteria weights and MCDM methods (after Choo et al. 1999)

MCDM method interpretation	Linking pin AHP	Referenced AHP	Saaty's AHP	Interactive MCDM	MVT SMART	Outranking ELECTRE
Marginal contribution	✓	✓	✓	✓	✓	×
Trade-offs	✓	✓	✓	✓	✓	×
Gradient of U(x)	✓	✓	✓	✓	✓	×
Scaling factor	✓	✓	✓	✓	✓	×
Linear coefficient	✓	✓	✓	✓	✓	×
Total partial value	×	✓	×	×	×	×
Typical partial value	✓	×	×	×	×	×
Discriminating power	×	×	×	×	✓	×
Binary vote value	×	×	×	×	×	✓
Partial value of x	×	×	×	✓	×	×
Searching parameter	×	×	×	✓	×	×
Criteria importance	✓	✓	✓	×	✓	✓
Question to DM	a	b	c	d	e, f	g

[a] Which is more important, the worth of x^1 (the linking alternative) in the criterion C_q or the worth of x^1 in the criterion C_k, and by how many times?
[b] Which is more important, the average worth of all the alternatives in the criterion C_q or the average worth of all the alternatives in the criterion C_k, and by how many times?
[c] Of the two criteria C_q and C_k being compared, which one is considered more important, and by how many times, with respect to the overall goal?
[d] With all scores in other criteria held constant, how much increase in the criterion C_q are you willing to accept as compensation for one unit loss in the criterion C_k?
[e] How much more important is criterion C_q than the least important criterion?
[f] What is the ratio of the contribution (to overall value) of the swing from worst to best in the criterion C_q to the contribution of the swing from worst to best in the criterion C_k?
[g] With all scores in other criteria held constant, is the increase in the criterion C_q by its preference threshold p_q of the greater, minor or the same importance compared with the increase in the criterion C_k by its preference threshold p_k?

2.9 Classification of Weighting Methods

In literature, different weighting methods have been proposed to assign weights to the criteria (Pöyhönen and Hämäläinen 2001; Stewart 1992). The simplest way to assign weights to criteria is 'equal weights method' that distributes weights equally among all the criteria. The 'Equal weights method' has been applied in many decision-making problems (Wang et al. 2009).

Weights assigned to criteria in multi-criteria evaluation method is an important step as final results of the multi-criteria decision-making method largely depend on such weights. Tervonen et al. (2009) state that assigning weights to criteria in a MCDM approach is the most difficult task. The main purpose of a weighting method is to attach cardinal or ordinal values to different criteria to indicate their relative importance in a multi-criteria decision-making method. These values are then used by the MCDM method in subsequent evaluation of the alternatives. A classification of weighting methods based on internal and external types of weighting methods is shown in Fig. 2.6.

Wang et al. (2009), classifies the rank-order method into three categories: subjective weighting method, objective weighting method and combination weighting method. The subjective methods determine criteria weights based on the preferences of the decision-makers. They explain the elicitation process more clearly and are the most used for MCDM in water resources management. They include SMART, AHP, SIMOS and the Delphi method. The objective weights are obtained by mathematical methods based on the analysis of initial data. The objective weight procedure is not very clear and includes methods such as least mean square (LMS), minmax deviation, entropy, TOPSIS and multi-objective optimization. The combination or optimal weighting methods are a hybrid of methods that include multiplication and additive synthesis.

There are also other weighting methods for assigning differential weights to decision criteria. These weighting methods can be divided into two categories: 'objective weighting methods' and 'subjective weighting methods.' In 'objective weighting methods,' weights are obtained by mathematical methods and decision-makers have no role in determining the relative importance of criteria (Wang et al. 2009). In the use of 'subjective weighting methods,' the process of assigning importance to criteria depends on the preferences of decision-makers, and has been more commonly used in different studies (e.g. Zardari 2008). On the basis of objective, subjective, and combined properties, classification of weighting methods is shown in Fig. 2.7.

2.9.1 *Subjective Weighting Methods*

In the subjective weighting methods, criteria weights are derived from the decision-maker's judgment on criteria. This means that the subjective methods are to

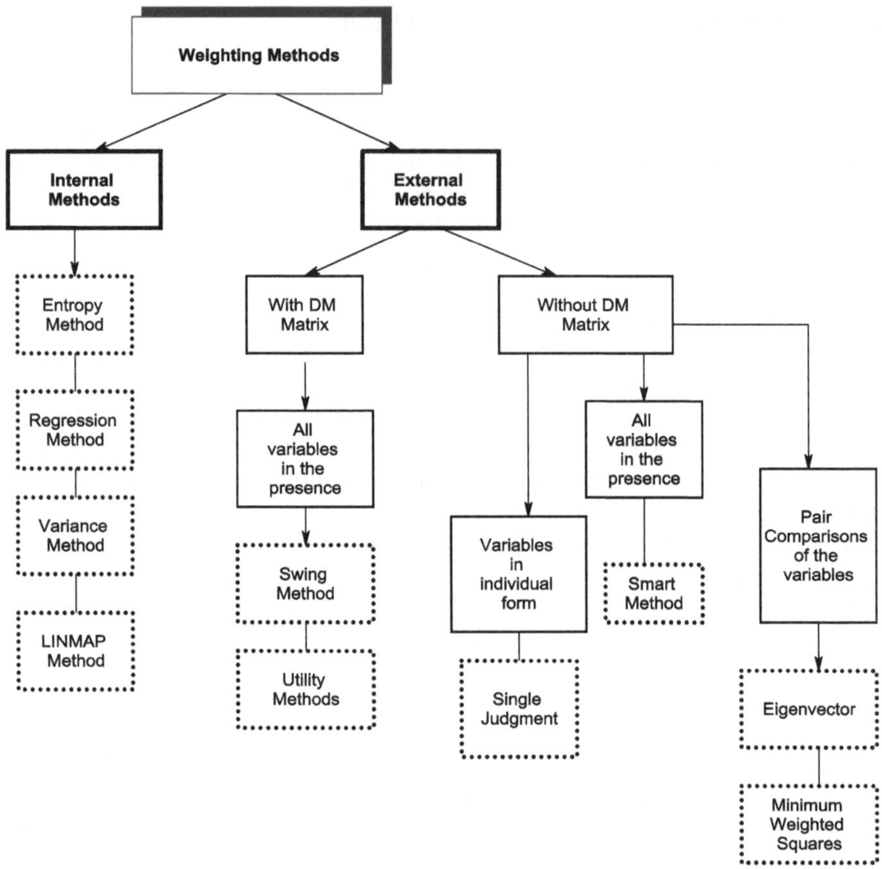

Fig. 2.6 Schematic diagram of the weighting methods

determine weights solely according to the preferences of decision makers. Criteria weights determined by the subjective weighting methods reflect the subjective judgment of the decision-maker, but analytical results or rankings of alternatives based on the weights can be influenced by the decision maker due to his/her level of knowledge and experience in the relevant field (Ahn 2011).

2.9.2 Objective Weighting Methods

In the objective weighting methods, preferences of decision maker on multiple criteria are not involved and the criteria weights are obtained from mathematical algorithms or models. The objective methods determine criteria weights by solving mathematical models automatically without any consideration of the decision

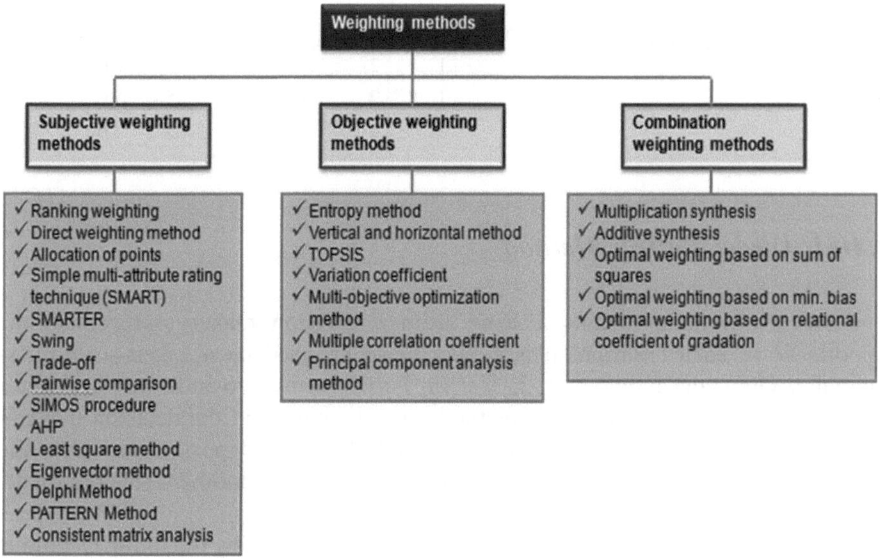

Fig. 2.7 Classification of weighting methods

maker's preferences. Objective weighting methods determine criteria weights by making use of the mathematical models, but they neglect the subjective judgment information of the decision maker (Aalianvari et al. 2012).

2.10 Popular Subjective Weighting Methods

The most popular weighting methods that have been used in previous multi-criteria decision-making studies are listed as below:

1. Direct Rating
2. Ranking Method
3. Point Allocation
4. Pairwise Comparison
5. Ratio Method
6. Swing Method
7. Graphical Weighting
8. Delphi Method
9. Simple multi-attribute ranking technique (SMART)
10. SIMOS Method

Table 2.4 Example of weighting using a rating scale

Criterion	Importance (1 = least, 7 = most)						
C_1	1	2	3	4	5	6	7
C_2	1	2	3	4	5	6	7
C_3	1	2	3	4	5	6	7

2.10.1 Direct Rating Method

The rating technique obtains a score from a decision maker to represent the importance of each criterion. It is similar to scales used on a Likert-scale questionnaire. Often the numbers 1–5, 1–7 or 1–10 are used to indicate importance (Nijkamp et al. 1990). The rating method does not constrain the decision maker's responses as the fixed point scoring method does. It is possible to alter the importance of one criterion without adjusting the weight of another. This represents an important difference between the two approaches.

An example of a survey task designed to elicit weights using the rating technique could ask a decision maker to show the importance of each criterion on an ordinal scale as shown in Table 2.4.

2.10.2 Ranking Method

The ranking method is the simplest approach for assigning weights to criteria. Essentially, the criteria are ranked in order from most important to least important. Once this is done, there are three main methods to calculate weights. They include:

1. rank sum,
2. rank reciprocal and
3. the rank exponent method (Malczewski 1999).

In rank sum, the rank position r_j is weighted and then normalized by the sum of all weights.

Rank reciprocal weights are derived from the normalized reciprocals of a criterions rank. The rank exponent method requires the decision maker to specify the weight of the most important criterion on a 0–1 scale. The value is then used in a numerical formula. To better understand how weights are calculated, the Table 2.5 is provided. It is based on the example given by Malczewski (1999). Again r_j is the rank of the criterion and n is the number of criteria.

These three ranking methods are very attractive due to their simplicity. They also provide a satisfactory approach to weight assessment. As a starting point in deriving weights, the three ranking methods provide a way to simplify multi-criteria analysis. However, they are limited by the number of criteria to be ranked. This method

Table 2.5 Ranking methods to assign weights (after Malczewski 1999)

		Rank sum		Rank reciprocal		Rank exponent	
		Weight	Normalized	Weight	Normalized	Weight	Normalized
Criterion	Straight rank	$(n - r_j + 1)$		$(1/r_j)$		$(n - r_j + 1)^p$ $p = 2$	
Agr.	4	2	0.133	0.250	0.109	4	0.073
Forests	2	4	0.267	0.500	0.219	16	0.291
Urban	5	1	0.067	0.200	0.088	1	0.018
Wetlands	1	5	0.333	1.000	0.438	25	0.454
Water	3	3	0.200	0.333	0.146	9	0.164
		15	1.000	2.283	1.000	55	1.000

Straight rank is first

1 = most important criterion

5 = least important of the five criterion

is really not appropriate for a large number of criteria since it becomes very difficult to straight rank as a first step. Another problem is in the lack of any real theoretical foundation. These techniques should be considered weight approximation techniques only (Malczewski 1999).

2.10.3 Point Allocation

In point allocation weighting method, the decision maker allocates numbers to describe the criteria weights directly. The decision maker is asked, for example, to divide 100 points among the criteria. In many experiments, the analysts do not fix the total number of points to be divided but the subjects are asked to give any numbers they liked to reflect the weights. The more points a criterion receives, the greater its relative importance. The total of all criterion weights must sum to 100. This method is easy to normalize. This is very easy weighting method. However, the weights obtained from the use of point allocation method are not very precise. This method could be difficult if the number of criteria increases to 6 or more.

Criteria that are important for ranking watersheds		Weight
	Elev change	10
	Wetlands	25
	Roads	5
	Row crops	40
	Mining	20

	Total	100

2.10.4 Pairwise Comparison Method

The pairwise comparison method is actually a very old psychometric technique that has been used by several generations of psychologists (Whitfield 1999). It is a well developed method of ordering criteria. Pairwise comparisons involve the comparison of each criterion against every other criterion in pairs. It can be effective because it forces the decision maker to give thorough consideration to all elements of a decision problem. The number of comparisons can be determined by:

$$o = \frac{m(m-1)}{2}$$

where:

o = the number of comparisons; and
m = the number of criteria

Calculating weights using the pairwise comparison method has three main steps (see Table 2.6). The first step is to develop a matrix comparing the criteria as shown in step one of Table 2.6. Next the intensity values are used to fill in the matrix of comparisons. Note that not all values need to be used. For example 1, 3, 5, 7, 9 or 1, 5, 9 could be used if the user finds it difficult to distinguish between definitions. With three criteria (price, slope, and view), the top right part of the matrix is filled in

Table 2.6 Pairwise comparison method weight calculation (after Strager 2002)

	Step 1			Step 2			Weights
	Price	Slope	View	Price	Slope	View	
Price	1	4	7	0.718	0.769	0.538	(0.718 + 0.769 + 0.538)/ 3 = 0.675
Slope	¼	1	5	0.179	0.192	0.385	(0.179 + 0.192 + 0.385)/ 3 = 0.252
View	1/7	1/5	1	0.102	0.039	0.077	(0.102 + 0.039 + 0.077)/ 3 = 0.073
	1.393	5.200	13.0	1.000	1.000	1.00	1.000

based on the comparisons. So for example, price is moderately to strongly preferred over the slope criterion therefore receiving a value of 4. The diagonal in the matrix is always 1 and the lower left values are inverse values because we make that assumption that the matrix is reciprocal. This completes the first step.

The second step is to compute the criterion weights. This is done by summing the values in each column, dividing each element by the column total, and dividing the sum of the normalized scores for each row by the number of criteria (3 criteria in this example).

The third step is to compute a consistency ratio. Many software programs such as Criterion Decision Plus and Expert Choice provide the consistency ratio for users. If the consistency ratio is less than 0.10, then the ratio indicates a reasonable level of consistency in the pairwise comparisons. If it is larger than 0.10, the values of the ratio are indicative of inconsistent judgments.

The pairwise comparison method is often criticized for simply asking for the relative importance of evaluation criteria without reference to the scales on which the criteria are measured. This fuzziness may mean that decision makers interpret the questions in different and possibly erroneous ways. Also, if many criteria are being compared, the number of individual comparisons may be cumbersome. Abbreviated pairwise comparisons can deal with this problem. Advantages of pairwise comparison include: the method requires only two criteria to be considered at one time, and the method has been tested theoretically and empirically for a variety of decision situations including spatial decision making (Malczewski 1999).

2.10.5 Ratio Weighting Method

The ratio method (Edwards 1977) requires the decision makers to first rank the relevant criteria according to their importance. The least important criterion is assigned a weight of 10 and all others are judged as multiples of 10. The resulting raw weights are then normalized to sum to one. The ratio method is an algebraic, decomposed, direct procedure.

2.10.6 Swing Weighting Method

The swing method (von Winterfeldt and Edwards 1986) starts from an alternative with the worst outcomes on all criteria or attributes. The decision maker is allowed to change one criterion from worst outcome to best. The decision maker is asked which 'swing' from the worst to the best outcome would result in the largest, second largest, etc., improvement. The criterion with the most preferred swing is most important, and given 100 points. The magnitudes of all other swing are expressed as percentages of the largest swing. Again, the derived percentages are the raw weights that are normalized to yield final weights. This method's strength is that it does take into account the range of each criterion, and it is a relatively simple, straightforward method. However, it does not allow participants to directly compare each criterion with each other.

2.10.7 Graphical Weighting Method

There are many variations on graphical weighting of criteria. One approach is to have a decision maker place a mark on a horizontal line. Criteria importance increases as the mark is placed further to the right end of the line. A quantitative weight can be calculated by measuring the distance from the mark to the left extremity of the line. Scores are usually normalized to obtain an overall weights vector. This approach enables decision makers to express preferences in a purely visual manner. The graphical weighting technique is sometimes criticized because it permits decision maker's to be carefree in assigning weights. For example, it is easy for a decision maker to place a mark on a horizontal line without considering the implications for criteria weights. In favor of graphical methods, however, is the ease and quickness with which they can be used. Many decision makers do not have sufficient time for some of the more complex and involved approaches (Hajkowicz et al. 2000). An example of graphical weighting method is shown in Fig. 2.8.

2.10.8 Delphi Method

In Delphi Methods the weights are derived in following three stages.

Stage 1: Participants are chosen. Initial data (what type of initial data is gathered?) is gathered and participants present their views on the policy.

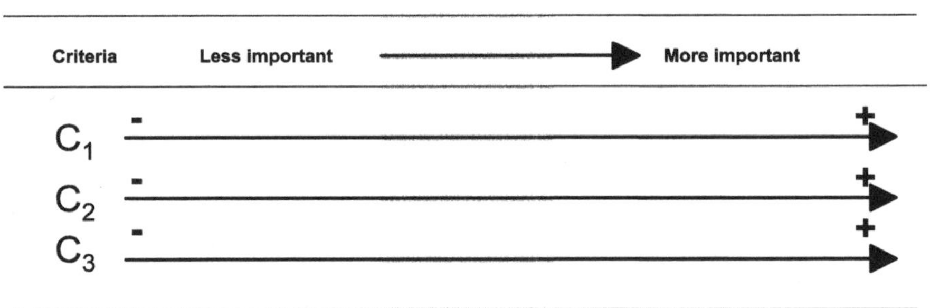

Fig. 2.8 Graphical weighting example

Stage 2: A list of possible alternatives is compiled and distributed to participants. Ideas are synthesized and a smaller number of possible policy recommendations are compiled.
Stage 3: An amended list of alternatives is distributed. These "policy" ideas are fine-tuned by the participants.

2.10.9 Simple Multi-attribute Rating Technique (SMART)

Simple multi-attribute rating technique (SMART) is originally described as the whole process of rating alternatives and weighting criteria by Von Winterfeldt and Edwards (1986). In this method decision maker is asked to rank the importance of criteria from worst levels to best levels. Then they assign 10 points to least important criteria, and an increasing number of points are assigned to the other criteria to address their importance relative to the least important criteria. The weights are calculated by normalizing the sum of the points to one. On this basis some new versions such as SMARTER and SMARTS presented to elicit the weights.

2.10.10 SIMOS Weighting Method

SIMOS (1990a, b) proposed a technique allowing any decision-maker (not necessarily familiarized with multi-criteria decision aiding) to think about and express the way in which he wishes to hierarchise the different criteria of a family F in a given context. This procedure also aims to communicate to the analyst the information he needs in order to attribute a numerical value to the weights of each criterion of F, when they are used in an ELECTRE type method (Roy and Mousseau 1996; Roy and Bouyssou 1993). The procedure has been applied to different real-life contexts; it proved to be very well accepted by decision-makers and we believe that the information obtained by this procedure is very significant from the decision-maker's preference point of view. However, the way SIMOS recommends to process the information needs a revision for two main reasons:

1. It is based on an unrealistic assumption. This occurs by the lack of essential information (as it was already underlined by Scharlig 1996).
2. It leads to process criteria having the same importance (i.e., the same weight) in a not robust way.

2.10.11 *Revised SIMOS Weighting Method*

Figueira and Roy (2002) developed a weighing. The main innovation of this procedure is relating a "playing card" to each criterion. The procedure can be summarized into four main steps as follows:

1. Each decision-maker (DM) is given n colored cards (or n criteria). Each card has the criterion name inscribed on it and objective of the criterion. A number of white cards (blank cards) are also provided.
2. The DM is then asked to rank the cards from the least important to the most important. If certain criteria are perceived to be of equal importance (same weighting), the cards are grouped together (same rank position).
3. The DMs are asked to insert the white cards between two successively ranked colored cards (or group of cards) in order to express their strong preference between criteria. The number of white cards is proportional to the difference between the importance of the considered criteria.
4. The DM is finally asked to answers the question "how many times more important the first ranked criterion (or group of criteria) is, relative to the last ranked criterion (or group of criteria)?"

2.10.12 *Fixed Point Scoring*

In this method the decision maker is required to distribute a fixed number of points amongst the criteria. A higher point score indicates that the criterion has greater importance. Often percentages are used as this is a measure with which many decision makers are familiar. The key advantage of fixed point scoring is that it forces decision makers to make trade-offs in a decision problem. Through fixed point scoring it is only possible to ascribe higher importance to one criterion by lowering the importance of another. This presents a difficult task to the decision maker which requires careful consideration of the relative importance of each criterion. Fixed point scoring is the most direct means of obtaining weighting information from the decision maker. It requires the least amount of operations to transform information supplied by the decision maker into a weights vector satisfying the requirements mentioned earlier.

2.11 Popular Objective Weighting Methods

Following is the list of popular objective weighting methods.

1. Entropy method.
2. Criteria Importance Through Inter-criteria Correlation (CRITIC).

3. Mean Weight.
4. Standard Deviation.
5. Statistical Variance Procedure.

2.11.1 Entropy Method

Entropy method is a measure of uncertainty in the information formulated using probability theory. It indicates that a broad distribution represents more uncertainty then the sharply peaked one (Deng et al. 2000). To calculate the weights by entropy method first the information matrix is normalized then following equations are used.

$$p_{ij} = \frac{x_{ij}}{\sum_{i=1}^{m} x_{ij}} \quad i = 1,\ldots,m;\ j = 1,\ldots,n$$

$$E_j = -\left(\sum_{i=1}^{m} p_{ij} \ln(p_{ij})\right) / \ln(m) \quad j = 1,\ldots,n$$

$$w_j = \frac{1 - E_j}{\sum_{i=1}^{n} (1 - E_k)} \quad j = 1,\ldots,n$$

where

x_{ij} = Original measured data
E_j = Information Entropy method
w_j = Entropy method Weight

2.11.2 CRITIC Weighting Method

In addition to the entropy method, any other method of measuring the divergence in performance ratings can be used to determine the criteria weights. Diakoulaki et al. (1995) have proposed the CRITIC (The Criteria Importance Through Intercriteria Correlation) method that uses correlation analysis to detect contrasts between criteria. First vector r_j of the normalized matrix is generated where r_j denotes the scores of all n alternatives.

Each vector r_j is characterized by the standard deviation (σ_j), which quantifies the contrast intensity of the corresponding criterion. So, the standard deviation of r_j is a measure of the value of that criterion to be considered in the decision-making process. Next, a symmetric matrix is constructed, with dimensions m × m and a generic element l_{jk}, which is the linear correlation coefficient between the vectors r_j

and r_k. It can be seen that the more discordant the scores of the alternatives in criteria j and k are, the lower is the value l_{jk}. In this sense, Eq. (2.1) represents a measure of the conflict created by criterion j with respect to the decision situation defined by the rest of the criteria:

$$\sum_{k=1}^{m} 1 - l_{jk} \tag{2.1}$$

The amount of information C_j conveyed by the jth criterion can be determined by composing the measures which quantify the above 2 notions through the multiplicative aggregation formula (Eq. 2.2). The higher the value C_j is, the larger is the amount of information transmitted by the corresponding criterion and the higher is its relative importance for the decision-making process. Objective weights are derived by normalizing these values to unity (Eq. 2.3).

$$C_j = \sigma_j \sum_{k=1}^{m} (1 - l_{kj}) \tag{2.2}$$

$$w_j = C_j \left[\sum_{k=1}^{m} C_k \right]^{-1} \tag{2.3}$$

r_j = Scores of all alternatives
C_j = Amount of information
w_j = Weight of Criteria

2.11.3 *Mean Weight (MW)*

In Mean Weight the weights are derived objectively by using equation $w_j = 1/n$, where n is number of criteria. This is based on the assumption that all criteria are of equal importance. Mean weight is used in MCDM when the there is no information from decision maker or information is not sufficient to reach a decision.

2.11.4 *Standard Deviation Method*

Standard deviation (SD) method is similar to Entropy method which assigns small weights to an attribute, if it has similar attribute values across alternatives. The SD method determines the weights of criteria in terms of their SDs through following equations (Jahan et al. 2012).

$$w_j = \sigma_j \Big/ \sum_{j=1}^{n} \sigma_j \quad j = 1, \ldots, n$$

$$\sigma_j = \sqrt{\frac{\sum_{i=1}^{m} \left(x_{ij} - \overline{x_j}\right)^2}{m}} \quad j = 1, \ldots, n$$

where

w_j = Weight of criteria
σ_j = Standard deviation

2.11.5 Statistical Variance Procedure

Statistical Variance procedure is an Objective Weighting method in which objective weights are derived. Initially statistical variance of information is calculated by

$$V_j = (1/n) \sum_{i=1}^{n} \left(x_{ij}^* - (x_{ij}^*)_{mean}\right)^2$$

V_j = Statistical Variance
X_{ij} = average value of data set of points

The objective weight can be obtained by following equation

$$w_j^o = \frac{V_j}{\sum_{i=1}^{m} V_j}$$

2.11.6 Integrated or Combined Weighting Methods

In Integrated or Combined Weighting methods the weights are derived from both subjective and objective information on criteria weights.

Weighting is a very critical task in decision making because it involves controversy and uncertainty (Chen et al. 2009) and it influences the final outcome, the ranking of alternatives. Several methods have been developed for this purpose, which are reported in the literature: swing weights; ranking; rating; pairwise comparison; trade-off analysis; qualitative translation, etc. Reviews of these methods are provided by Beinat (1997), Malczewski (1999) and Sharifi et al. (2004). Crucial factors for selecting the most appropriate method for assigning weights to criteria for a certain decision problem are the number of criteria and the grade of uniqueness between them. Two factors were taken into account when making the decision to choose methods for the Evaluation module. First, the number of criteria involved in the evaluation process carried out in this model is quite small, i.e. five. This falls within the so called ‘seven plus or minus two’ range that is considered as the maximum number of entities that can be simultaneously processed by the human brain (Miller 1956). Second, given that certain evaluation criteria are explicit in terms of their context and meaning, it was judged that two of the most straightforward and popular methods should be utilized and provided by the Evaluation module: direct ranking and qualitative rating. The direct ranking method allows the user to enter weights when they are known *a priori* or developed using another method while the qualitative ranking method developed here offers users a way of developing the weights within the process. These two methods are explained in more detail below.

2.11.7 Direct Ranking

Direct ranking (or direct estimation) is the most straightforward method for assigning values to criteria (that sum up to 1) when the number of criteria is small and manageable. However, even for such a small number of criteria, it is not straightforward when weighting values have two or more decimals. For instance, sometimes is not easy to justify why a criterion has a weight of 0.2 and another criterion has 0.18; it is even more difficult to differentiate a criterion from another by assigning weights of 0.125 and 0.120. Thus, weighting with this method can be reliable and accurate when values have one decimal, i.e. 0.1, 0.2 or two decimals with the last digit being 5, i.e. 0.15 or 0.25. Because these preconditions cannot always be fulfilled, we provide the planner with a modified rating method called ‘qualitative rating’, which has been proposed in this research and is explained in the next section.

2.11.8 Qualitative Rating Method

Ranking methods involve the ordering of criteria to identify the most important to the least important criteria or *vice versa*. Several procedures (e.g. rank sum, rank reciprocal and rank exponent method) are then utilized for estimating a numerical value of weights based on that rank order (Malczewski 1999). Although these methods are simple, they involve a great disadvantage since they do not provide the potential to rank two or more criteria with equal importance, a fact that is obviously not reasonable in practice.

Kamal (2012) has presented a comprehensive review on recent literature on applications of various weighting methods in different fields of research. Tables 2.7, 2.8 and 2.9 show subjective, objective, and combined weighting methods applications along with the list of countries in which these studies were completed.

2.12 Objective Weighting Methods Used in Past Studies

See Table 2.8.

2.13 Subjective and Objective Weighting Methods Used in Past Studies

See Table 2.9.

2.14 Selection of Weighting Method

Hobbs (1980) states that different weighting methods produce different set of criteria weights and final results of the multi-criteria decision-making methods are sensitive to criteria weights. Therefore, it is paramount to emphasis on the selection of weighting method for solving a multi-criteria decision problem.

The selection of a particular method is highly dependent on the particular decision problem (Hobbs 1980; Zardari et al. 2010).

Hajkowicz et al. (2000) applied five weighting methods to weight six economic, environmental and social criteria. Comparisons were made between criteria weights obtained from each method and decision makers evaluated each method for its ease of use and how much it helped clarify the decision problem.

Findings of their study indicate that, in general, decision makers will assign similar weight values to the criteria with the different methods. However, minor variations in weight vectors such as these have the potential to cause significant changes in the subsequent ranking of alternatives. This indicates that it is undesirable to rely upon any single weighting technique in a MCDM approach as there may be bias associated with that particular technique (Fig. 2.9).

There are several methods for transforming experts' judgments into relative weights. Eckenrode (1965) found no significant differences among the techniques they investigated. Since no method is clearly superior, the preferred method in any application depends on the intended use of the scale (ordinal, interval, or ratio level of measurement), the time required to use it, the subjects' mental attitudes, their understanding of the overall problem, their perception of the instructions for weighting the criteria, and their understanding of the criteria definitions.

Table 2.7 Application of subjective weighting methods in past studies (after Kamal 2012)

Reference	MCDM method	Subjective weighting method	Area of research	Publication	Country
Machiwal et al. (2011)	AHP	Pairwise comparison, eigenvector method	Ground water potential zones	Water Resources Management	India
Calizaya et al. (2010)	AHP	Pairwise comparison	Integrated water resources management	Water Resources Management	Bolivia
Kodikara et al. (2010)	PROMETHEE	SIMOS procedure	Urban water supply	European Journal of Operational Research	Australia
Özcan et al. (2011)	TOPSIS, ELECTRE, grey theory	SIMOS procedure	Ware house selection	Expert Systems with Applications	Turkey
Tervonen et al. (2009)	ELECTRE Tri, SMAA Tri	SIMOS procedure	Sorting problems	European Journal of Operational Research	Portugal, Finland
Shanian et al. (2008)	ELECTRE Tri	Revised SIMOS procedure	Material selection	Knowledge-Based Systems	Canada, USA
Chen and Zhang (2010)	TOPSIS	Pairwise comparison	Bid evaluation	Conference Paper	China
Supriyasilp et al. (2009)	AHP	Pairwise comparison	Hydropower development	Energy Policy	Thailand
Guo and Yu (2011)	TOPSIS, AHP	Pairwise comparison	Evaluation of railway	Conference Paper	China
Myllyviita et al. (2012)	MADM	ReCiPe LCI assessment method, SMART	EIA of biomass production	Journal of Cleaner Production	Finland
Yeh and Chang (2009)	Fuzzy group MCDM	Pairwise comparison	Modeling of subjective evaluation	European Journal of Operational Research	Australia, Taiwan
Balasubramaniam et al. (2007)	ELECTRE III, weighted summation	SWING	Selection of remediation techniques for petroleum contaminated land	Chemosphere	UK

(continued)

Table 2.7 (continued)

Reference	MCDM method	Subjective weighting method	Area of research	Publication	Country
Hämäläinen and Alaja (2008)	MCDM	SWING	Weighting biasness in environmental decision analysis	Ecological Economics	Finland
Morais and Almeida (2012)	PROMETHEE II	Ranking	Water resources management	Omega	Brazil
Chang and Yeh (2001)	MADM	SAW, SP, TOPSIS	Evaluation of airlines	Omega	Taiwan, Australia
Arnette et al. (2010)	VAHP	Pairwise comparison	Ranking of watershed goals	Environmental Modelling and Software	USA
Gallego-Ayala and Juízo (2011)	AHP, A'WOT analysis	Pairwise comparison	Implementation of IWRM	Physics and Chemistry of the Earth, Parts A/B/C	Mozambique
Chen et al. (2011)	FMCDM, FAHP	Pairwise comparison	Selection of best environment-watershed plan	Applied Soft Computing	Taiwan
Sorvari and Seppälä (2010)	DST based on MAVT	Ratio method, pairwise comparison	Risk management options for contaminated sites	Science of the Total Environment	Finland
Wang et al. (2002)	DSSDSS based on MAUT	SMART	Dewatering	Building and Environment	Singapore
Gómez-Limón et al. (2003)	MAUF, EUT	Additive utility function	Agricultural risk aversion	European Journal of Operational Research	Spain
Chowdhury and Rahman (2008)	MCDA	Weighted sum method	Water resources management	International Journal of Environmental Science and Technology	Australia, Bangladesh

(continued)

Table 2.7 (continued)

Reference	MCDM method	Subjective weighting method	Area of research	Publication	Country
Jaber and Mohsen (2001)	AHP	Pairwise comparison	Water resources supply	Desalination	Jordan
Afshar et al. (2011)	FMCDM based on TOPSIS	Fuzzy UNEP, pairwise comparison	River basin	Water Resources Management	Iran
Toosi and Samani (2012)	AHP, SAW	Pairwise comparison, SAW	Water pollution control	Water International	Iran
Eskandari et al. (2012)	SAW	Pairwise comparison, rank ordering method	Landfill siting	Waste Management	Iran
Delle and Filippi (2009)	MAVT	Ratio weighting, pairwise comparison, trade-off	Transport assessment projects	European Transport Research Review	Italy
Pietersen (2006)	MCDA	SMART, SWING	Management of ground water resources	Water SA	South Africa
Yu and Hu (2010)	Fuzzy TOPSIS	Voting method, rating method	Evaluation of manufacturing plants	Computers and Industrial Engineering	Taiwan
Yang et al. (2008)	ANP, TOPSIS	Pairwise comparison	Vendor selection	Information Sciences	India
Yang et al. (2010)	AHP	Pairwise comparison	Energy efficiency assessment	Energy Policy	China
Abrishamchi et al. (2005)	AHP	Pairwise comparison	Water transfer project	Water Resources Management	Iran
Chou et al. (2008)	FSAW	Fuzzy numbers	Selection of facility location	European Journal of Operational Research	Taiwan
Meyar-Naimi and Vaez-Zadeh (2012)	AHP	Pairwise comparison	Energy policy making	Energy Policy	Iran

(continued)

Table 2.7 (continued)

Reference	MCDM method	Subjective weighting method	Area of research	Publication	Country
Yanhui et al. (2012)	AHP	Pairwise comparison	Water resources vulnerability	Procedia Earth and Planetary Science	China
Gallego-Ayala (2012)	AHP, modified TOPSIS	Pairwise comparison	Irrigation water pricing	Mathematical and Computer Modeling	Mozambique
Wang et al. (2012)	Fuzzy comprehensive evaluative, AHP	Pairwise comparison	Risk assessment of floor water	International Journal of Rock Mechanics and Mining Sciences	China
Sargaonkar et al. (2011)	AHP	Pairwise comparison	Artificial ground water recharge	Environmental Earth Sciences	India
Kaya and Kahraman (2011)	Fuzzy VIKOR, AHP	Fuzzy pairwise comparison	Selection of forested water shed	Expert Systems with Applications	Turkey
Anagnostopoulos and Petalas (2011)	Fuzzy multi criteria cost-benefit approach, FAHP	Fuzzy weights, pairwise comparison	Evaluation of irrigation projects	Agricultural Water Management	Greece
Wu et al. (2012)	Fuzzy integrated assessment	Delphi, pairwise comparison	Evaluation of ecological quality	Marine Pollution Bulletin	China
Schuwirth et al. (2012)	MAVT, MAUT	SWING, trade-off	Pharmaceutical removal from hospital wastewater	European Journal of Operational Research	Switzerland
Hayashi (2000)	Goal programming	SMART	Agricultural resources management	European Journal of Operational Research	Japan
Kim et al. (2010)	MCDA	Delphi method	Contaminated sediments	Integrated Environmental Assessment and Management	South Korea
Peacock et al. (2007)	MAUT	SWING	Setting health service priorities using PBMA	Social Science and Medicine	Canada, Australia

(continued)

Table 2.7 (continued)

Reference	MCDM method	Subjective weighting method	Area of research	Publication	Country
Bryan (2010)	MCDA	Pairwise comparison, SWING	Management of natural capital and ecosystem services	Biological Conservation	France
Montazar and Behbahani (2007)	AHP	Pairwise comparison	Selection of irrigation system	Bio Systems Engineering	Iran
Zhang (2009)	AHP	Pairwise comparison	Allocation of water conservation	Water Resources Management	China
Benzerra et al. (2012)	Decision support approach	Pairwise comparison	Urban drainage system management	Journal of Environmental Management	France, Algeria
Aalianvari et al. (2010)	AHP, Fuzzy Delphi method	Delphi method. pairwise comparison, fuzzy weights	Potential of ground water flow	Arabian Journal of Geosciences	Iran
Montazar and Gaffari (2012)	AHP	Pairwise comparison	Crop planning within irrigation command	Irrigation and Drainage	Iran
Dai and Cai (2012)	Fuzzy comprehensive evaluation	Pairwise comparison	Ecological environment quality evaluation	Advanced Materials Research	China
Gu et al. (2012)	FAHP	Pairwise comparison	Evaluation of water quality	Advanced Materials Research	China
Chen et al. (2010)	MAUT	SMART	Construction method selection	Automation in Construction	China, USA
Huang (2011)	MAUT	SMARTER	Recommended system for e-Commerce	Electronic Commerce Research and Applications	Taiwan
Nieto-Morote et al. (2011)	FAHP	Pairwise comparison, fuzzy numbers	Selection of a tri generation system	International Journal of Energy Research	Spain
Rouyendegh and Erkan (2012)	Fuzzy ELECTRE, FAHP	Pairwise comparison	Academic staff selection	Human Factors and Ergonomics in Manufacturing	Turkey

(continued)

Table 2.7 (continued)

Reference	MCDM method	Subjective weighting method	Area of research	Publication	Country
Ferretti (2011)	Spatial ANP	Pairwise comparison	Development of landfill siting	Journal of Multi-Criteria Decision Analysis	Italy
Al-Juaidi et al. (2010)	Goal programming	Weighted sum method	Analysis of treated waste water for agriculture purposes	JAWRA Journal of the American Water Resources Association	Palestine
Wang and Lee (2009)	Fuzzy recognition model	Iterative weights integrated method, linguistic variables	Flood control operations	JAWRA Journal of the American Water Resources Association	China
Gómez-Limón and Riesgo (2004)	MAUT	Additive utility function	Irrigation water pricing	Agricultural Economics	Spain
Panagopoulos et al. (2012)	AHP	Pairwise comparison	Urban water demand	Water Resources Management	Greece
Latinopoulos (2009)	MOP	SE-ENV	Efficient water and land resources allocation	Environment, Development and Sustainability	Greece
Opricovic (2009)	VIKOR method	VIKOR method	Water resources planning	Water Resources Management	Serbia
Straton et al. (2011)	Compromise programming	Double weighting method	Use of scenario development and Water planning	Water Resources Management	Australia
Bobylev (2009)	AHP	Pairwise comparison	Environmental security, impacts on groundwater in urban areas	Decision Support for Natural Disasters (BOOK)	Russia

(continued)

Table 2.7 (continued)

Reference	MCDM method	Subjective weighting method	Area of research	Publication	Country
Chung and Lee (2009a)	CP, ELECTRE II, regime method, and evamix method	Pairwise comparison	Identification of hydrological vulnerability	Water Resources Management	South Korea
Chung and Lee (2009b)	Composite programming, ELECTRE II, evamix method, and regime method	Pairwise comparison	Prioritization of water management for sustainability	Journal of Environmental Management	South Korea
Kenyon (2007)	A participant-led multi-criteria approach	Rank reversal, ROC	Flood risk management	Ecological Economics	UK
Peniwati and Brenner (2008)	AHP	Pairwise comparison	Policies for drinking water	European Journal of Operational Research	Indonesia
Srdjevic (2007)	MCDM, AHP	Preference schedule, the hare system, plurality voting, The Borda count, approval voting, pairwise comparisons voting	Group decision making in water resources management	Decision Support Systems	Serbia
Xevi and Khan (2005)	Goal programming	Double weighting mechanism	Water management	Journal of Environmental Management	Australia
Karimi et al. (2011)	AHP, FAHP	Pairwise comparisons	Selection of wastewater treatment process	International Journal of Environmental Science and Technology	Iran
Chang et al. (2010)	FAHP	Pairwise comparisons, fuzzy numbers	Development of the metropolitan water availability index	Journal of Environmental Management	USA
Simonovic and Verma (2008)	Fuzzy Pareto optimal set	Fuzzy numbers	Water resources multi criteria decision making under uncertainty	Physics and Chemistry of the Earth	India

(continued)

Table 2.7 (continued)

Reference	MCDM method	Subjective weighting method	Area of research	Publication	Country
Mutikanga et al. (2011)	PROMETHEE	Revised SIMOS	Water loss management	Water Resources Management	Netherland, USA
De Feo and De Gisi (2010)	AHP	Priority scale to use pairwise comparisons	Selection of municipal solid waste site	Waste management	Italy
Lee and Chan (2008)	AHP	Pairwise comparisons	Assessment of urban renewal proposals	Social Indicators Research	China
Ramjeawon and Beerachee (2008)	AHP	Pairwise comparisons	Site selection of sanitary landfills	Waste Management and Research	Mauritius
Chen et al. (2009)	FAHP	Pairwise comparisons, fuzzy numbers	Planning the environment watershed	Cutting-Edge Research Topics on Multiple Criteria Decision Making	Taiwan
Bana e Costa et al. (2004)	MACBETCH approach	Simple aggregation function	Evaluation of flood control measures	Water Resources Management	Portugal
Qureshi and Harrison (2003)	AHP	Pairwise comparisons	Riparian vegetation policy options	Small-Scale Forestry	Australia
Joubert et al. (2003)	MCDA	ROC, SWINGS	Evaluation of water supply augmentation and water demand management options	Journal of Multi-Criteria Decision Analysis	South Africa
Prato and Herath (2007)	Additive utility multiple criteria function	Fixed point scoring, judgment analysis, pairwise comparison	Catchment management	Ecological economics	USA, Australia
Chuntian and Chau (2002)	Fuzzy pattern recognition	Fuzzy weights	Reservoir flood control	European Journal of Operational Research	China
Zarghami et al. (2008)	Compromise programming	Simple additive weighting	Urban water management	Water Resources Management	Iran

(continued)

Table 2.7 (continued)

Reference	MCDM method	Subjective weighting method	Area of research	Publication	Country
Abrishamchi et al. (2005)	Compromise programming	Double weighting mechanism, pairwise comparison	Urban water supply	Journal of Water Resources Planning and Management-Asce	Iran
Alipour et al. (2010)	FMCDM	Fuzzy weights	Water diversion	Expert Systems with Applications	Iran
Taylor and Ryder (2003)	Delphi Method	Delphi survey	Water reservoir management	Journal of the American Water Resources Association	USA
Qureshi and Harrison (2001)	Weighted summation, expected value, EVAMIX	Pairwise comparison	Riparian re-vegetation options	Journal of Environmental Management	Australia
Fontana et al. (2011)	SMARTER	Revised SIMOS procedure	Urban water conservation	Evolutionary Multi-Criterion Optimization. (BOOK)	Brazil
Morais and Almeida (2010)	Group decision making approach	Ranking method	Water network rehabilitation	Water SA	Brazil
Morais and Almeida (2006)	ELECTRE I	Pairwise comparison	Investment in water supply systems	Water SA	Brazil
Morais and Almeida (2007)	PROMETHEE V	Pairwise comparison	Leakage management	Resources, Conservation and Recycling	Brazil
Duarte et al. (2009)	GISEQ	n-point scale, pairwise comparisons, n-points scale modified (complemented with a ranking)	Ranking of water supplier performance	Energy, Environment, Ecosystems, Development and Landscape Architecture	Portugal

(continued)

Table 2.7 (continued)

Reference	MCDM method	Subjective weighting method	Area of research	Publication	Country
Delgado-Galván et al. (2010)	AHP	Pairwise comparisons	Water leakage management	Mathematical and Computer Modeling	Spain
Bello-Dambatta et al. (2009)	AHP	Pairwise comparisons	Contaminated land management	Advanced Engineering Informatics	UK
Liu et al. (2006)	TOPSIS	Rating method	Wetland catchment assessment	Environmental Management	Australia
Manoliadis et al. (2007)	AHP, compromise programming	Pairwise comparisons, trade-off analysis	Site selection for construction temporary facilities	Operational Research	Greece
Montazar and Zadbagher (2010)	AHP	Pairwise comparisons	Assessing global water productivity of irrigation networks	Water Resources Management	Iran
Olson (2004)	TOPSIS	Equal weights, regression analysis, ordinal rank	Comparison of weights	Mathematical and Computer Modeling	USA
Fattahi and Fayyaz (2010)	Compromise programming	Pairwise comparison	Water management	Water Resources Management	Iran
Biswas et al. (2012)	AHP	Pairwise comparison, rating method, ranking method	Water shed management	Environmental Monitoring and Assessment	Bangladesh, USA
Lamy et al. (2002)	SAW	Weighted sum	Watershed management	JAWRA Journal of the American Water Resources Association	USA
Prato (1999)	Utility theory, surrogate worth tradeoff,	Fixed-point scoring, paired comparison, judgment analysis	Watershed management	Natural Resource Modeling	Australia

(continued)

Table 2.7 (continued)

Reference	MCDM method	Subjective weighting method	Area of research	Publication	Country
	free iterative search, stochastic dominance				
Peng and Zhou (2011)	FMOWRMM, WSM, LM	Paired comparison, fuzzy numbers	Water allocation plan	Water and Environment Journal	China
Zardari et al. (2010)	ELECTRE I, II	Conjoint analysis	Water allocation	JAWRA Journal of the American Water Resources Association	Pakistan
Levy et al. (2007)	MCDSS	Pairwise comparison	Flood hazard mitigation and emergency response in urban watersheds	JAWRA Journal of the American Water Resources Association	
Babaei et al. (2012)	AHP, TOPSIS	Pairwise comparison	Water shortage	Water and Environment Journal	Iran

Table 2.8 Application of objective weighting methods in past studies (after Kamal 2012)

References	MCDM method	Objective weighting method	Area of research	Publication	Country
Shemshadi et al. (2011)	Fuzzy VIKOR	Entropy method	Supplier selection	Expert Systems with Applications	Iran
García et al. (2010)	Goal programming GP, TOPSIS	CRITIC	Ranking of firms	Computers and Operations Research	Spain
Kabak and Uelengin (2010)	Fuzzy logic	Entropy method	Supply chain and production model	Journal of Universal Computer Science	Turkey
Wang and Luo (2010)	MCDM	Entropy method, SD, CRITIC, Ma et al.'s method, CCSD	Economic benefit assessment	Mathematical and Computer Modeling	China
García et al. (2010)	Goal programming, TOPSIS	Entropy method, weighted euclidean distances	Ranking of banks	Mathematical and Computer Modelling	Spain
Zou et al. (2006)	Fuzzy synthetics	Entropy method	Water quality assessment	Journal of Environmental Sciences	China
Yurdusev and O'Connell (2005)	Compromise programming	Ideal point method	Water resources planning	Water Resources Management	Turkey, UK
Wang and Zhan (2012)	Dynamic multi-criteria decision making model	Entropy method	Bid evaluation	Systems Engineering Procedia	China
Zhang et al. (2011)	TOPSIS	Entropy method	Evaluate the tourism destination competitiveness	Tourism Management	China
Liu et al. (2010)	Fuzzy comprehensive evaluation	Entropy method	Water quality assessment	Expert Systems with Applications	China

(continued)

Table 2.8 (continued)

References	MCDM method	Objective weighting method	Area of research	Publication	Country
Srdjevic et al. (2004)	TOPSIS, Goal programming, modified TOPSIS	Entropy method	Water management	Water Resources Management	Brazil
Xiang (2008)	Fuzzy similarity	Entropy method	Selection of wetland and points for water quality monitoring points	Computer And Computing Technologies In Agriculture, Volume II	China
Singh (2000)	MCDM	Entropy method	Modelling and decision tool for environment and water resources	Water SA	USA
Jia et al. (2011)	Fuzzy comprehensive evaluation	Entropy method	Adjust peak value and frequency in power system	Intelligent Computing and Information Science	China
Wang and Men (2007)	TOPSIS	Variant coefficient	Classification of water resources	Proceedings of the 2007 International Conference on Agriculture Engineering	China
Xue et al. (2008)	TOPSIS	Variant coefficient method	Evaluation customer service satisfaction	Conference Paper	China

Table 2.9 Subjective and objective weighting methods used in past studies (after Kamal 2012)

References	MCDM method	Subjective weighting method	Objective weighting method	Area of research	Publication	Country
Ding (2011)	Fuzzy TOPSIS	Fuzzy pairwise comparison	Entropy method	Ranking of alternatives	Journal of Marine Science and Technology-Taiwan	Taiwan
Jiang et al. (2011)	GRA	Pairwise comparison	Entropy method	Flood disaster	Conference Paper	China
Harmancioglu et al. (2008)	Simple additive weighting, comprise programming, TOPSIS	Pairwise comparison	Entropy method, CRITIC method, SD, MW	Water Resources Management	Desalination	Turkey
Wang et al. (2008)	Fuzzy multiple criteria model	Pairwise comparison on	Entropy method	Selection of cool storage	Energy and Buildings	China
Rao and Patel (2010)	Proposed MADM	Pairwise comparison, point allocation method, digital logic method	Statistical variance procedure	Material selection	Materials and Design	India
Rao et al. (2011)	MADM	Pairwise comparison, point method	Statistical variance procedure	Industrial robot selection	Robotics and Autonomous Systems	India, Thailand

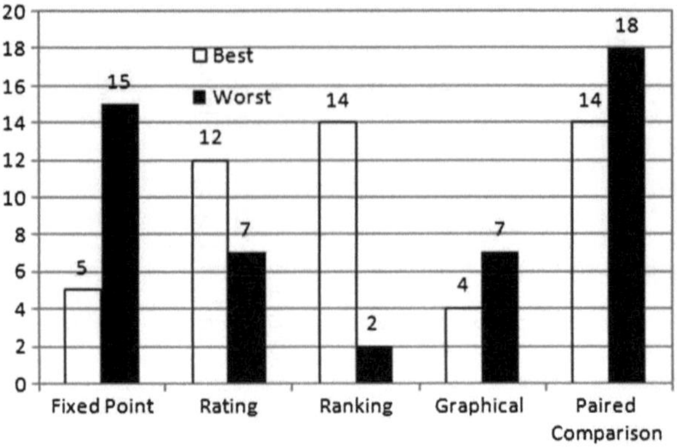

Fig. 2.9 Number of decision makers identifying weighting methods as the 'best' or 'worst'. The 'best' and 'worst' categories do not total to 55 due to non-responses from some decision makers (after Hajkowicz et al. 2000)

Methods for choosing weights have been surveyed before (Eckenrode 1965; Huber 1974) but not with an eye to the theoretical requirements imposed by weighting summation. A number of techniques are presented below, as are their applications in power plant siting. In general, methods that ask decision makers to choose weights directly do not guarantee that the weights are theoretically valid. Methods that derive weights by ensuring that the decision rule is consistent with trade-offs expressed by decision makers are more likely to yield valid weights. Those methods, however, generally are more difficult to apply.

In the absence of test/retest data, a rough indication of the reliability of the weights may be obtained by comparing the weights chosen by the two persons most likely to have similar values and knowledge. The two Maryland participants have worked together on siting studies and have similar perspectives on the siting problem. The correlation between their rating weights was 0.783, the highest between any pair of rating weight sets. Their indifference trade-off weight sets had a correlation of 0.624. These two correlations can be taken as a measure of method reliability. Both are much higher than the mean "between method" correlations. Hence, choice of weighting method appears to result in greater differences between weight sets than what can be attributed to method unreliability alone.

These large differences in candidate area sets contradict assertions made by some researchers (Dawes and Corrigan 1974; Wainer 1976) to the effect that weighting is unimportant as similar rank orders will often result from very different weight sets. Choice of weights is important here because siting is concerned only with the best few alternatives, not the entire rank order. Correlations of suitability are high, but the candidate areas differ.

Nijkamp et al. (1990) suggest various methods to estimate criteria weighting. These are broadly divided into two main approaches: direct and indirect estimation.

Direct estimation of criterion weights refers to the expression of relative importance of the objectives or criteria in a direct way through questionnaire surveys. Respondents are asked questions within which their priority statements are conveyed in numerical terms. Respondents can be members of the design team, representatives from the client, local council and public (Seabrroke et al. 1997). This is another opportunity for the increasing demand for public participation in the decision-making process (Joubert et al. 1997).

Direct estimation method techniques come in various forms:

- The trade-off method where the decision-maker is asked directly to place weights on a set of criteria to all pairwise combination of one criterion with respect to all other criteria.
- The rating method where the decision-maker is asked to distribute a given number of points among a set of criteria to reflect their level importance.
- The ranking method where the decision-maker is asked to rank a set of criteria in order of their importance.
- The seven-points (or five-points) scale which helps to transform verbal statements into numerical values.
- The paired comparison, which is similar to the seven-point scale, obtains the relative importance of criteria by comparing all pairs of criteria on a non-points scale.

However, all these methods run into trouble when the number of objectives becomes large (van Pelt 1993; Hobbs and Meier 2000). When this happens, objectives may have to be structured in a hierarchical model to separate objectives into different levels (Saaty 1994).

The indirect approach is based on investigating the actual behavior of respondents in the past. Weights are obtained through estimating actual previous behavior derived from ranking alternatives or through an interactive procedure of obtaining weights by questioning the decision-maker and other involved parties. Hypothetical weights may also be used in some projects. Here, the analyst prepares weights to represent the opinion of specific groups in the community, then policy-makers may comment accordingly. Each approach has restrictions and limitations in terms of accuracy and cost. Their usefulness strongly depends on the time required and the attitude of respondents (Voogd 1983; Nijkamp et al. 1990; Hobbs and Meier 2000).

Hobbs (1980) used two weighting methods, one deriving weights from trade-offs made by decision makers and the other asking decision makers to choose weights on a scale of 0–10. They found that the power plant locations picked by the two weightings methods were differing noticeably. This shows that different weighting methods produce different set of criteria weights and final results of the multi-criteria decision-making models are sensitive to criteria weights. Therefore, it is paramount to emphasis on the selection of weighting method for solving a multi-criteria decision problem.

Weighting of criteria is subjective and has direct influence on the results of prioritizing strategy options. It is therefore critical that criteria weights are determined rationally and truthfully (Hobbs 1980).

2.15 Weighting Methods Supported by Softwares

Kamal (2012) has presented a detailed survey of literature and observed that a large number of softwares are developed for the multi-criteria decision-making methods and but not separate softwares were developed for the weighting methods. He stated, however, different MCDM methods support various weighting methods. Some softwares support only one weighting method and some support more than one weighting method. A review on softwares' availability for various MCDM methods supported for different weighting methods is given as below (Kamal 2012).

2.15.1 Pairwise Comparison

Thirty-two multi-criteria decision-making methods softwares were found which supported pairwise comparison weighting method. These are:

1. Super Decisions
2. V.I.S.A
3. 1000 Mind
4. Decision Lens
5. Make it Rational
6. Expert Choice
7. D-Sight
8. Decision Plus
9. DEMATEL
10. Criterium Decision Plus SELECT PRO SOFTWARE LLC
11. DEFINITE
12. AHP (Analytic Hierarchy Process) Calculation software by CGI
13. USAGE: Calculation (Software): Weights of AHP
14. MULINO-DSS software
15. HIPRE 3
16. Web-HIPRE
17. Joint Gain
18. Logical Decisions Portfolio v6.2
19. Logical Decisions for Windows v6.2
20. Mind Decider
21. Select Best Voll version
22. Open Decision Maker
23. Right Choice DSS
24. Select Pro
25. AHP project
26. AHP with Qualica planning Suit
27. AHP Decision
28. Choice Results

29. M-AHP
30. Vanguard System
31. MVLSoft Very Good Choice
32. ANOVA

2.15.2 Point Allocation Method

Two softwares which support point allocation weighting method found in the literature. These are:

1. 1000 Mind
2. QUALIFLEX

2.15.3 Ranking Method

Four softwares were supporting ranking method. These are:

1. Logical Decision
2. QUALIFLEX
3. Select Pro
4. RPM decision

2.15.4 Rating Method

Three softwares were supporting rating weighting method. These softwares are:

1. 1000 Mind
2. Criterium DecisionPlus
3. MULINO-DSS software

2.15.5 SMART Weighting Method

From detail review of literature and internet browsing, Kamal (2012) found seven softwares which were supporting SMART weighting method and these include:

1. 1000 Mind
2. Criterium® Decision Plus

3. HIPRE 3
4. Win Pre
5. Web-HIPRE
6. Equity3
7. MACBETH

2.15.6 SWING Weighting Method

A large number of commercial softwares were available in the literature that were found supporting SWING weighting method. These include:

1. V.I.S.A
2. 1000 Mind
3. Logical Decision
4. Win Pre
5. Web-HIPRE
6. Tree Top
7. RICH Decisions
8. Promax 2010 Standard
9. QMms (Quantitative Methods for Management Science)
10. Logical Decisions Portfolio v6.2
11. Logical Decisions for Windows v6.2
12. MindDecider
13. IDS (Intelligent Decision System)
14. Hiview3
15. Equity3
16. Analytica 4.2
17. RISK

2.15.7 Trade-off Weighting Method

Only two softwares were available in literature that was supporting trade-off weighting method, which are:

1. MULINO-DSS software
2. Criterium Decision Plus

2.15.8 Delphi Method

For Delphi weighting method, only one software was found in the literature, which is:

1. Delphi Decision Aid

2.15.9 Revised SIMOS Procedure

Revised SIMOS weighting method was supported by only one software which is:

1. SRF software

Kamal (2012) summarized all available softwares for each weighting method. Table 2.10 shows how many softwares were available for some specific popular weighting methods.

Based on the availability of number of softwares, Kamal (2012) provided ranking of weighting methods (popularity of weighting methods) as follows:

Pairwise Comparison > SWING > SMART > Ranking method > Rating Method > Trade-Off = Point Allocation > Delphi Method = Revised SIMOS Procedure (where '>' is 'better than').

2.16 Advantages and Disadvantages of Weighting Methods

Kamal (2012) summarizes advantages and disadvantages of some popular weighting methods. The advantages and disadvantages are presented as below for the each weighting method.

Table 2.10 Distribution of number of softwares against each weighting method

Weighting method	Number of softwares
Delphi method	1
Direct weighting method	5
Pairwise comparison	32
Point allocation method	2
Ranking method	4
Rating method	3
Revised SIMOS procedure	1
SMART	7
SWING	17
Trade-off	2

2.16.1 Pairwise Comparison

Advantages

1. Pairwise comparison is useful when the decision maker is unable to rank the alternatives holistically and directly with respect to a criterion.
2. This method is easy to calculate. The results are clear, and especially distinctive for issues about qualitative factors which are used for decision making or evaluation.
3. The Pairwise comparisons method is often used as an intermediate step in multi-criteria decision making, when the decision maker (DM) is unable to directly assign criteria weights or scores of alternatives.
4. Pairwise comparison can be effective because it forces the decision makers to give through consideration to all elements of decision problem (Hajkowicz et al. 2000).
5. Pairwise comparison is commonly used to estimate preference values of finite alternatives with respect to a given criterion.
6. In the pairwise comparison prioritization process it is also assumed that DMs are able to express the strength of their preferences by providing additional cardinal information.
7. This methodology is found to be the most user transparent and scientifically sound methodology for assigning weights representing the relative importance of criteria.

Disadvantages

1. Pairwise comparisons suffer from two major shortcomings. First, many do not allow participants to explicitly convey a sense of distance in their choices, since participants are usually asked to simply select an attribute from a pair. Second, the complexity of comparing items in pairs can be quite high for large attribute sets, usually resulting in conflicting choices and lack of transitivity.
2. There is inconsistency at DM's idea in pairwise comparison and it increases either by higher number of attributes or judging the important degree.
3. The main difficulty is to reconcile the inevitable inconsistency of the pairwise comparison matrix elicited from the decision makers in real-world applications.
4. The Pairwise comparison is used to derive weights for Analytical hierarchy process. As the size (n) of the hierarchy increases, the number of Pairwise comparisons increases rapidly. The completion of $n(n-1)/2$ comparisons (quite high in realistic problems) can become a very difficult task for the decision maker when applied to all levels of the hierarchy.
5. The pairwise comparison is seems to be insufficient and imprecise to capture the right judgments of decision-maker(s) with vagueness and uncertainty of data.
6. In many real world applications, human pairwise judgment is highly ambiguous and uncertain.
7. Pairwise comparison is an important step in AHP to be completed by the experts. However, AHP is widely criticized for such tedious process especially

when a large number of criteria or alternatives are involved. Someone may doubt the expert judgments because people are very likely to feel tired and lose patience during this process and therefore, they may not make their judgments conscientiously. They may change their minds frequently in order to ascertain the acceptance of the consistency ratio (CR) value as well as shorten the whole process. To avoid such drawback, only reasonable and manageable amounts of criteria are contained in the model and the author of this study has acted as a facilitator to take over the judgment process (Lee and Chan 2008).

2.16.2 Simple Multi-attribute Rating Technique (SMART)

Advantages

1. The Simple Multi-attribute Rating Technique (SMART) can be used to quickly obtain a total weighted score (Huang 2011).
2. SMART is one of the most applicable MCDM methods, and since the majority of the panelists' were not familiar with MCDM methods, the method had to be simple (Yeh and Chang 2009).
3. SMART method is easy to modify when the number of impact categories increased (Yeh and Chang 2009).
4. The SMART approach utilizes ratio-scales to assess panellists' preferences (Yeh and Chang 2009).
5. SMART is a useful technique since it is simple, straightforward and requires less time in decision making that is quite important for those involved in the decision-making process (Gu et al. 2012).
6. In SMART, changing the number of alternatives will not change the decision scores of the original alternatives and this is useful when new alternatives are added (Chen and Hou 2004; Panagopoulos et al. 2012).
7. Using SMART in performance measures can be a better alternative than other methods (Gu et al. 2012).
8. The SMART is popular because its analysis incorporates a wide variety of quantitative and qualitative criteria (Chen and Hou 2004).
9. SMART has been successfully applied in MCDM problems, this approach is ineffective when dealing with the inherent imprecision of linguistic valuation in the decision-making (Gu et al. 2012; Chen and Hou 2004).
10. The advantage of the smart model is that it is independent of the alternatives (Panagopoulos et al. 2012; Afshar et al. 2011).
11. The nontechnical participants especially felt that SMART was easier to understand as compared to Trade-off method (Dai et al. 2012).

Disadvantages

1. It has been stressed that the comparison of the importance of the attributes is meaningless, if it does not reflect the consequence ranges of the attributes as well (Von Winterfeldt and Edwards 1986).
2. One of the limitations of this technique is that it ignores the interrelationships between parameters (Demirci et al. 2009).
3. The ratings of alternatives are not relative; changing the number of alternatives considered will not in itself change the decision scores of the original alternatives (Valiris et al. 2005).
4. Due to the large number of attributes, we determined that the SMARTS method would be too difficult to implement and defend (Benzerra et al. 2012).

2.16.3 Point Allocation Method

Advantages

1. The key advantage of fixed point scoring/Point Allocation Method is that it forces decision makers to make trade-offs in a decision problem (Deng et al. 2000).
2. Through fixed point scoring/Point Allocation (PA) Method it is only possible to ascribe higher importance to one criterion by lowering the importance of another. Fixed point scoring/Point Allocation Method is the most direct means of obtaining weighting information from the decision maker (Deng et al. 2000).
3. It requires the least amount of operations to transform information supplied by the decision maker into a weights vector satisfying the requirements mentioned earlier (Deng et al. 2000).
4. According to using a simple PA system or other technique probably works well with a small number of attributes (Mustajoki et al. 200).
5. The weights elicited by Point allocation method were more reliable than those elicited by direct rating in a test-retest situation (Von Winterfeldt and Edwards 1986).

Disadvantages

1. The point allocation method/Fixed point scoring is a more difficult task since it is easier to take 100 as the weight for the most important attribute and then to allocate weights relative to this 100 starting point as the weight of successive attributes. The decision maker has no need to worry about the constraint that the total must be some specified value. Since the set of cognitive operations required to use the two methods is different, there is every possibility that different decision weights will result (Demirci et al. 2009).

2. Although this method of determining weights and the direct rating method would seem to be minor variants of each other, however, in practice they produce different profiles of decision weights (Demirci et al. 2009).
3. It is a difficult task for the decision maker to ascribe higher importance to one criterion by lowering the importance of another which requires careful consideration of the relative importance of each criterion (Deng et al. 2000).

2.16.4 Revised SIMOS' Procedure

Advantages

1. The revised SIMOS' procedure is simple and easy to use, requiring little computational effort, thus increasing its applicability (Fontana et al. 2011).
2. It is shown to be efficient when evaluating alternatives on qualitative attributes when applying an additive method (Fontana et al. 2011).
3. Revised SIMOS' procedure to minimize the rounding off errors when the normalized criteria weights are calculated (Fontana et al. 2011).
4. The 'Revised SIMOS' Procedure', the technique used to collect information on weights, proved to be well accepted by all participants (Özcan et al. 2011).
5. SIMOS' technique allows any DM (not necessarily familiarized with multi criteria decision aiding) to think about and express the way in which he wishes to hierarchies the different criteria of a family "F" in a given context. This procedure also aims to communicate to the analyst the information he needs in order to attribute a numerical value to the weights of each criterion of "F".
6. The procedure has been applied to different real-life contexts; it proved to be very well accepted by DMs and we believe that the information obtained by this procedure is very significant from the DM's preference point of view.
7. The software developed allows not only an easy collection of different data sets but also a quick processing of the information thus obtained.
8. In multi criteria decision aiding contexts, the new procedure and the software can also be used to adapt or convert a scale of a given criterion into an interval scale or a ratio scale.

Disadvantages

1. In Revised SIMOS' procedure, interval scale evaluation is required (Fontana et al. 2011).
2. In cases where the DM's spontaneous response to the question 'How many times more important is the most important PM (or group of PMs), relative to the least important PM (or group of PMs)?' differs substantially from the total number of cards used (including blank cards), the calculated normalized weights of PMs shows a distortion of the original PM rank order expressed by the DM (Özcan et al. 2011).

3. It is suggested that identify whether the DM's understanding of this scale (with blank cards inserted) would be a ranking order or a ratio scale (Özcan et al. 2011).
4. It can occasionally process some criteria that have the same importance in an uncontrolled manner (Chen and Zhang 2010).
5. Similar to the AHP method, there is no reference made to criteria scales and therefore certain combinations of weights may be excluded (Chen and Zhang 2010).

2.16.5 Trade-off Weighting Method

Advantages

1. Its advantage is the strong theoretical foundation (Taylor and Ryder 2003).
2. Trade-off method does not require a person to assign weights to, nor state relative importance of, the attributes or criteria directly. Instead, it asks one to state how much compromise he is willing to make between two attributes or criteria when an ideal combination of the two is not attainable.
3. Some weighting methods derive weights from trade-offs decision makers are willing to make. Such weights are likely to be theoretically valid (Hajkowicz et al. 2000).
4. A common feature of AHP and SMART methods is that they rely on ratio comparisons about the "relative importance" of attributes, although the resulting weights are not explicitly linked to unit changes in the component value functions. To avoid this shortcoming, several authors have recommended the use of the trade-off method (Delgado-Galván et al. 2010).

Disadvantages

1. In practice, the trade-off method is difficult and time consuming to use compared with the other methods (Fatthi and Fayyaz 2010).
2. The trade-off method was considered more difficult and some participants had real problems understanding the underlying logic behind it (Chang et al. 2010).
3. The elicitation of these exact weights imposes a level of precision that is often absent in people's minds (Morais and Almeida 2010).
4. DM may find difficulty in giving precise responses to the trade-off questions (Delgado-Galván et al. 2010).

2.16.6 Delphi Method

Advantages

1. A key advantage of the approach is that it avoids direct confrontation of the experts (Afshar et al. 2011).
2. A benefit to theory building derives from asking experts to justify their reasoning (Afshar et al. 2011).
3. Delphi researchers employ this method primarily in cases where judgmental information is indispensable, and typically use a series of questionnaires interspersed with controlled opinion feedback (Afshar et al. 2011).
4. The Delphi method, a consensus-building tool, is a promising process to promote and encourage involvement from all stakeholders during the evaluation framing process (Dai et al. 2012).
5. The Delphi method removes geographic challenges and time boundaries allowing all stakeholders to participate (Dai et al. 2012).
6. The most important advantage of the Delphi method is that it leads to a group decision. Group decisions have many merits such as the avoidance of the extreme judgment of individual assessors (Chou et al. 2008).

Disadvantages

1. It has some limitations including the potential of falling victim to the band wagon effect. Dominant personalities can unduly influence the face-to-face group (Anagnostopoulos and Petalas 2011).
2. Critics have noted other limitations of the Delphi methodology: potential for sloppy execution, crudely designed questionnaires, poor choice of panelists, unreliable result analysis, limited value of feedback and consensus, and instability of responses among consecutive Delphi rounds.
3. A further limitation, fatigue, occurs when there are a large number of topics or questions per Delphi topic, or when questions are difficult to understand (Peng and Zhou 2011).
4. It consumes high cost for conducting operation (Chou et al. 2008).
5. The drawback of the Delphi method is that it is very time-consuming and expensive due to more than one round being needed (Chou et al. 2008).

2.16.7 SWING Method

Advantages

1. Swing Weight Matrix was very useful to assess, explain, and defend weights. The Swing Weight Matrix Method provided an efficient and effective means to discuss, assess, brief, and explain the attribute weights (Von Winterfeldt and Edwards 1986).

2. We believe this method has four advantages over traditional weighting methods. First, it develops an explicit definition of importance. Second, it forces explicit consideration of the variation of measures. Third, it provides a framework for consistent swing weight assessments. Fourth, it provides a simple yet effective framework to present and justify the weighting decisions (Von Winterfeldt and Edwards 1986).
3. Swing method overcomes many of the problems of constant-sum ratio estimation; is relatively simple, transparent and easy to use; and produces weights which are practically indistinguishable from indifference methods (Demirci et al. 2009).
4. The swing weights technique is more parsimonious than techniques that involve Pairwise comparisons like AHP when many (>4) criteria need to be weighted (Valiris et al. 2005).
5. The Swing method uses a reference state in which all attributes are at their worst level and let the interviewee assign points to states in which one attribute moves to the best state. The weights are then proportional to these points (Hayashi 2000).
6. It is fairly fast and interviewees readily give answers (Hayashi 2000).
7. Another advantage of the Swing method is that it does not depend on the shape of the value functions of the sub-objectives. Only the attribute ranges must be known and the levels of the best and worst outcomes (in most cases corresponding to the endpoints of the ranges). This makes it possible to elicit weights prior to assessment of the value functions of the sub-objectives, which can reduce the splitting bias, as mentioned below (Hayashi 2000).
8. The subjects of this study found the swing weighting method relatively easy to follow, although most participants indicated that they would have preferred further explanation (Hämäläinen and Alaja 2008).
9. The SWING method is of intermediate complexity and was found by participants to be relatively easy to use, making its employment in questionnaire survey appropriate (Hämäläinen and Alaja 2008).

Disadvantages

1. Swing method holds the risk that people respond without thoroughly considering the consequences of their answers (Hayashi 2000).
2. The disadvantages are that the technique is based on direct rating, it does not include consistency checks, and the extreme outcomes to be compared may not correspond to a realistic alternative, which makes the questions difficult to answer.
3. In terms of external validity, assessed by comparing the participant weights with weights externally elicited from experts by comparing with Tradeoff method, the Trade-off method performed better than swing weighting (Hämäläinen and Alaja 2008).

2.16.8 Entropy Method

Advantages

1. Entropy method can computes unbiased relative criteria weights, entropy approach enables measuring the source and determining the relative weights of criteria (w_1, w_2, …, w_m) in rather simple and straightforward manner (Srdjevic et al. 2004).
2. Entropy approach has been proved as sufficiently reliable in identifying both contrast intensity and conflict of criteria and computing their weights appropriately (Srdjevic et al. 2004).
3. It suggests if the available information is adequate or not and if not, then additional information should be sought. In this way it brings the model, the modeller, and the decision-maker closer (Singh 2000).
4. It permits a quantitative assessment of efficiency and benefit/cost parameters (Singh 2000).
5. The entropy method for determination of weight considers adequately the information of values all the monitoring sections provided to balance the relationship among numerous evaluating objects. This weakens the bad effect from some abnormal values and makes the result of evaluation more accurate and reasonable.
6. The entropy method for determination of weight is a very effective method for evaluating indicators.
7. The traditional entropy method focuses on the discrimination among data to determine attribute weights. If an attribute can discriminate the data more effectively, it is given a higher weight.
8. The Entropy method produces more divergent coefficient values for all the criteria. We regard this phenomenon as favourable to the Entropy method as it can better resolve the inherent conflict between the criteria embedded in Multi-attribute decision problems (Diakoulaki et al. 1995).

Disadvantages

1. Its possible disadvantage is related to proper problem sizing, i.e. preserving that the decision matrix contains sufficiently large set of alternatives (Srdjevic et al. 2004).
2. It does not seem that considering the weights only based on entropy values without expert judgment would be sufficient.
3. The weights of attributes determined by the Correlation coefficient and Standard deviation (CCSD) method are more comprehensive and convincing than entropy weights. The former considers not only the amount of information each attribute contains, but also the impact of each attribute on decision making; while the latter takes no account of the mutual relationships among attributes (Mustajoki et al. 2004).
4. The entropy technique does not give scope to designer's preferences.

5. General purpose MADM techniques, such as entropy, could not effectively model public sector university ranking decision problems. Such decision problems require a new methodology to be developed.

2.16.9 Rank Ordering Centroid

Advantages

1. The key advantage of ROC Methodology is its simplicity in surveying.
2. ROC is simple and easy to follow (Chang et al. 2010).
3. ROC weights represent excellent trade- offs between ease of assessment and efficacy of selection of the best or near best alternative (Morais and Almeida 2010).
4. ROC weights possess other attractive properties. The best ROC alternative has the highest average value over the entire weight simplex, and ROC is the expected value of the weight distribution consistent with the information. Of greater usefulness is the fact that ROC is a specific example of Centroid weights (CW), which generalizes to any convex weight set specified by linear equalities in the unspecified weights (Morais and Almeida 2010).
5. This method is a simple way of giving weight to a number of items ranked according to their importance. The decision makers usually can rank items much more easily than give weight to them.
6. The four methods (RS, RR, ROC, and EW) are compared by using a simulation study and report that the ROC weights appear to perform better than the other approximate weights. They have also shown that the ROC weights are given by the arithmetic mean of the extreme points of ranked weights (Morais and Almeida 2010).
7. A common conclusion of these studies is that ROC weights have an appealing theoretical rationale and appear to perform better than the other rank-based schemes in terms of choice accuracy.

Disadvantages

1. The weights which are given by ROC are highly dispersed (Chang et al. 2010).

2.16.10 CRITIC Method

Advantages

1. The weights derived incorporate both contrast intensity and conflict which are contained in the structure of the decision problem (Jahan et al. 2012).

2. The method developed is based on the analytical investigation of the evaluation matrix for extracting all information contained in the evaluation criteria (Jahan et al. 2012).
3. The method can be easily converted into an algorithmic form (Jahan et al. 2012).
4. The weights derived from the method CRITIC proposed in this paper are found to embody the information which is transmitted from all the criteria participating in the multi-criteria problem. In addition objective weights offer an insight into the nature of the dilemmas created by the existence of conflicting criteria and enable the incorporation of interdependent criteria (Jahan et al. 2012).

Disadvantages

1. CCSD method such as no specific requirement of normalization formulations, clearer modeling mechanism than the CRITIC method (Mustajoki et al. 2004).

Chapter 3
Research Methodology and Results

3.1 Introduction

In this chapter we have provided details on the survey completed with postgraduate and undergraduate students registered in course Water Resources Management (MAL1023) and Integrated Water Resources Management (SAB4613) in Semester 1, Session 2012 at Faculty of Civil Engineering, Universiti Teknologi Malaysia. The aim of the survey was to know which weighting method was more simple in administration and was producing relatively more reliable set of criteria weights. Here we stress that the sample size for conducting the survey was limited and may not satisfy some sampling standards and guidelines. Therefore, the results of the survey are only applicable for the project scope and objectives and beyond this the proposed methodology can be tested with larger number of survey participants where it is likely that the different sets of criteria weights may be obtained. The details of the survey are presented in the subsequent sections of this chapter.

3.2 Methodology

In order to demonstrate weighting procedure and to get different sets of criteria weights from the use of ranking and rating weighting methods, we hypothesized a watershed for which management plan was to be developed. Watershed management criteria were selected from literature and these criteria were short listed based on these yard-sticks: (1) the selected criterion must change in time and space; (2) it should be measurable; (3) it should be understandable to all stakeholders; (4) the criterion should be relevant to watershed management; (5) the criterion should be simple to implement (if needed to do so), etc. After short-listing the watershed

N.H. Zardari et al., *Weighting Methods and their Effects on Multi-Criteria Decision Making Model Outcomes in Water Resources Management*,
SpringerBriefs in Water Science and Technology, DOI 10.1007/978-3-319-12586-2_3

Table 3.1 Short-listed watershed management criteria that were used in a survey

Criteria	Watershed management criteria
1	Provide good water quality
2	Provide good habitat for fish and aquatic species
3	Provide economic development incentives
4	Provide adequate flood control
5	Provide high quality of life by increasing green space
6	Provide adequate protection from droughts
7	Controlling soil erosion
8	Protect natural/cultural heritage and forests
9	Promote recreational activities
10	Protection of sensitive lands (e.g. flood plains and valley lands)
11	Efficient use of resources (land and water)
12	Balancing ecological, economic and social interests of watershed
13	Improve water supply reliabilities to inhabitants
14	Improve nutrient management
15	Protect woodlands, wetlands and habitats
16	Protect life and property from natural hazards
17	Safeguarding rivers, lakes and streams
18	Improve groundwater and surface water quantity and quality

management criteria, it was planned to develop priority ranking of these criteria based on the stakeholders' preferences. We state that this project was not aim to solve any real-world problem, rather to focus on developing a comprehensive and logical-based methodology for eliciting weights of watershed management criteria. Later the proposed methodology may be used to solve some real-world watershed and water resources problems. Table 3.1 provides list of eighteen short-listed watershed management criteria which we used in survey for developing their priority ranking from stakeholders' responses.

Postgraduate and undergraduate students of Faculty of Civil Engineering, Universiti Teknologi Malaysia participated in the survey. The survey questionnaire was designed in a way where biasness was eliminated or minimized. We split watershed management criteria into three groups and changed their order in the questionnaire for different groups of respondents.

A procedure for obtaining weights for the selected watershed management criteria is shown in a schematic diagram depicted in Fig. 3.1.

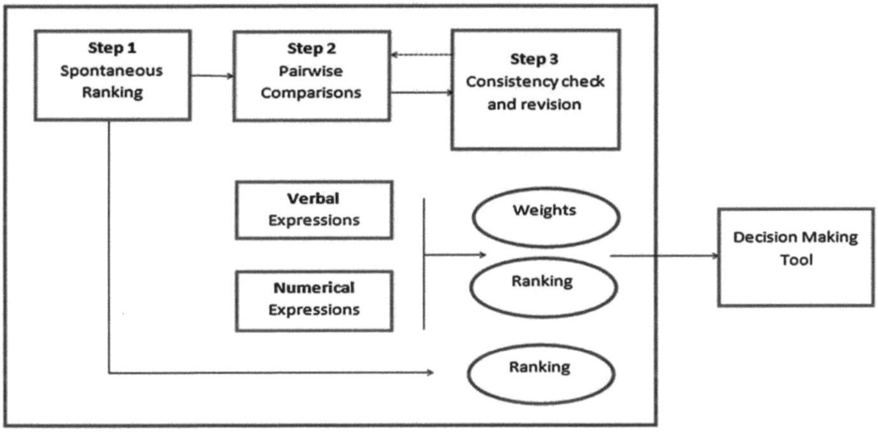

Fig. 3.1 Proposed weighting methodology

3.3 Survey Questionnaire

The survey questionnaire was consisted of five questions. In the first question, the survey participants were asked to allocate points to different criteria based on their importance in watershed management. For example in one set of questions the survey participants were asked to choose 'the most important' criterion first and give rating 100 to it. The rest of criteria can be given 0–99 points according to their importance in watershed management plan development. The full questionnaire is shown below along with the statements at the beginning of each question.

Question-1 Statement
Please identify the ***MOST important*** watershed management criterion from the list below and give it a rating of 100. Next, rate each of the other criteria in turn on a scale of 0–99, to indicate how important they would be compared to the MOST important criterion in YOUR decision. A high rating score means that the criterion would be important.

Criteria code	Watershed management criteria	Rating
C-1	Provide good water quality	
C-2	Provide good habitat for fish and aquatic species	
C-3	Provide economic development incentives	
C-4	Provide adequate flood control	
C-5	Provide high quality of life by increasing green space	

(continued)

(continued)

Criteria code	Watershed management criteria	Rating
C-6	Provide adequate protection from droughts	
C-13	Improve water supply reliabilities to inhabitants	
C-14	Improve nutrient management	
C-15	Protect woodlands, wetlands and habitats	
C-16	Protect life and property from natural hazards	
C-17	Safeguarding rivers, lakes and streams	
C-18	Improve groundwater and surface water quantity and quality	

Question-2 Statement

Please identify the ***LEAST important*** criterion from the list below and give it a rating of 10. Next, consider each of the other criteria in turn and indicate how important they would be compared to the LEAST important criterion in YOUR decision. All ratings should be greater than 10. (For example, if X were 15 times as important as Y, the least important criterion, you would give X a rating of 150).

Criteria code	Watershed management criteria	Rating
C-1	Provide good water quality	
C-2	Provide good habitat for fish and aquatic species	
C-3	Provide economic development incentives	
C-4	Provide adequate flood control	
C-5	Provide high quality of life by increasing green space	
C-6	Provide adequate protection from droughts	
C-13	Improve water supply reliabilities to inhabitants	
C-14	Improve nutrient management	
C-15	Protect woodlands, wetlands and habitats	
C-16	Protect life and property from natural hazards	
C-17	Safeguarding rivers, lakes and streams	
C-18	Improve groundwater and surface water quantity and quality	

Question-3 Statement

Please identify the ***MOST important*** watershed management criterion from the list below and give it a rating of 100. Next, rate each of the other criteria in turn on a scale of 0–99, to indicate how important they would be compared to the MOST important criterion in YOUR decision. A high rating score means that the criterion would be important.

Criteria code	Watershed management criteria	Rating
C-1	Provide good water quality	
C-2	Provide good habitat for fish and aquatic species	
C-3	Provide economic development incentives	
C-4	Provide adequate flood control	
C-5	Provide high quality of life by increasing green space	
C-6	Provide adequate protection from droughts	
C-7	Controlling soil erosion	
C-8	Protect natural/cultural heritage and forests	
C-9	Promote recreational activities	
C-10	Protection of sensitive lands (e.g. flood plains and valley lands)	
C-11	Efficient use of resources (land and water)	
C-12	Balancing ecological, economic and social interests of watershed	

Question-4 Statement

Please identify the ***LEAST important*** criterion from the list below and give it a rating of 10. Next, consider each of the other criteria in turn and indicate how important they would be compared to the LEAST important criterion in YOUR decision. All ratings should be greater than 10. (For example, if X were 15 times as important as Y, the least important criterion, you would give X a rating of 150).

Criteria code	Watershed management criteria	Rating
C-1	Provide good water quality	
C-2	Provide good habitat for fish and aquatic species	
C-3	Provide economic development incentives	
C-4	Provide adequate flood control	
C-5	Provide high quality of life by increasing green space	
C-6	Provide adequate protection from droughts	
C-7	Controlling soil erosion	
C-8	Protect natural/cultural heritage and forests	
C-9	Promote recreational activities	
C-10	Protection of sensitive lands (e.g. flood plains and valley lands)	
C-11	Efficient use of resources (land and water)	
C-12	Balancing ecological, economic and social interests of watershed	

Question-5 Statement
Please assign rank-order to the following watershed management targets/criteria according to their importance in managing watersheds. Most important criterion should be assigned with Rank-1 and the second most important criterion with Rank-2 and so on. The least important criterion should be given Rank-18.

Criteria code	Watershed management criteria	Rank
C-1	Provide good water quality	
C-2	Provide good habitat for fish and aquatic species	
C-3	Provide economic development incentives	
C-4	Provide adequate flood control	
C-5	Provide high quality of life by increasing green space	
C-6	Provide adequate protection from droughts	
C-7	Controlling soil erosion	
C-8	Protect natural/cultural heritage and forests	
C-9	Promote recreational activities	
C-10	Protection of sensitive lands (e.g. flood plains and valley lands)	
C-11	Efficient use of resources (land and water)	
C-12	Balancing ecological, economic and social interests of watershed	
C-13	Improve water supply reliabilities to inhabitants	
C-14	Improve nutrient management	
C-15	Protect woodlands, wetlands and habitats	
C-16	Protect life and property from natural hazards	
C-17	Safeguarding rivers, lakes and streams	
C-18	Improve groundwater and surface water quantity and quality	

3.4 Questionnaire Administration

We administered the questionnaire to two different groups of students. One group was consisted of postgraduate (Masters by Coursework) students registered in Water Resources Management course (MAL1023) and other group was consisted of undergraduate students (Final Year Civil Engineering) registered in course Integrated Water Resources Management (SAB4613). The following section provides analysis of survey data collected from 15 students of postgraduate students (MAL1023). Data analysis of undergraduate students (SAB4613) is given in the subsequent sections of this chapter.

3.4.1 Postgraduate Survey Data Analysis

A total of 15 postgraduate students were asked to participate in the survey and fill the questionnaire. The project leader explained the objectives of the survey to the survey participants well before conducting the survey. All participants were explained about each and every watershed management criteria included in the survey questionnaire. They were asked whether they understood the questionnaire. Once all agreed on the contents and meaning of the questionnaire and watershed management criteria, the participants were asked to complete the questionnaire. They were given 5 min for each of the five questions included in the questionnaire. This survey was completed on 5th April, 2012. The survey data analysis and results for the postgraduate students group is given as below. Responses of the postgraduate students and undergraduate students on five questions are shown in Appendix A. As an example, we have presented only two responses one from postgraduate group and one from undergraduate group (Tables 3.2 and 3.3).

Preferences of all postgraduate and undergraduate students on all watershed management criteria for all five questions of the questionnaire are given in Appendix B. However, as an example, the preferences of respondents for only Question-1 of the questionnaire are given in Tables 3.4 and 3.5.

Table 3.2 Postgraduate 1 (PG1) preferences on five questions of the survey questionnaire

Criteria code	Question-1	Question-2	Question-3	Question-4	Question-5
C-1	90	120	80	120	4
C-2	80	100	70	80	12
C-3	70	50	50	60	15
C-4	50	50	50	50	17
C-5	50	30	60	50	16
C-6	50	10	50	10	18
C-7			80	100	11
C-8			90	80	13
C-9			70	60	14
C-10			90	150	6
C-11			100	150	2
C-12			90	150	7
C-13	60	90			8
C-14	60	90			9
C-15	100	150			1
C-16	80	120			10
C-17	90	120			3
C-18	70	100			5

Table 3.3 Undergraduate 1 (UG1) preferences on five questions of the survey questionnaire

Criteria code	Question-1	Question-2	Question-3	Question-4	Question-5
C-1	100	200	100	200	1
C-2	70	80	53	10	15
C-3	79	120	68	50	11
C-4	78	111	64	16	13
C-5	72	90	70	32	12
C-6	77	110	63	15	14
C-7			99	180	3
C-8			98	160	4
C-9			73	78	10
C-10			84	90	8
C-11			92	120	5
C-12			80	81	9
C-13	84	150			6
C-14	82	145			7
C-15	65	20			16
C-16	59	10			18
C-17	63	15			17
C-18	98	180			2

3.5 Weights for the Watershed Management Criteria

In the survey questionnaire the respondents were asked to show their preferences for all eighteen watershed management criteria by using five different approaches. In Question-1, the respondents were asked to assign 100 points to the most important criterion and the rest of criteria may be given points 1–99 according to their importance in watershed management plan. We evaluated the survey responses of postgraduate and undergraduate groups in combination of all five questions as well as for each question separately. These weights were later aggregated to get overall weights of all watershed management criteria.

Similarly, respondents' preferences on different watershed management criteria were evaluated for Questions-2, 3, 4 and 5.

It can be seen from criteria weights shown in Tables 3.6, 3.7, 3.8, 3.9, 3.10, 3.11, 3.12, 3.13, 3.14 and 3.15 that some weighting methods were producing criteria weights with a large range (difference between maximum criterion weight and minimum criterion weight) and other producing weights in narrow range. For example, in Question-1 where survey participants were asked to show their preferences on watershed management criteria by allocating 100 points to the 'most important' criterion and assign points between 1 and 99 for the second most important criterion and so on. It was found that this type of weighting method produced criteria weights which were very close and in a narrow range which making difficult for decision-makers to choose the most important criteria for

Table 3.4 Preferences of postgraduate group on all 18 criteria for Question-1

Criteria code	Question-1 (most important and max 100 points and others 1–99 points)														
	PG1	PG2	PG3	PG4	PG5	PG6	PG7	PG8	PG9	PG 10	PG 11	PG 12	PG 13	PG 14	PG 15
C-1	90	100	100	100	100	100	100	45	80	100	100	70	85	80	100
C-2	80	95	80	85	85	60	65	28	55	40	2	40	70	90	90
C-3	70	40	80	50	30	90	80	80	50	50	12	55	90	50	70
C-4	50	70	70	75	50	50	75	20	70	70	60	80	100	60	95
C-5	50	60	65	60	99	70	92	70	85	30	20	40	65	10	85
C-6	50	50	10	70	60	20	85	30	60	65	70	65	80	30	75
C-13	60	75	95	65	80	5	76	50	75	80	10	85	60	40	90
C-14	60	55	70	60	90	15	45	35	45	45	1	40	20	75	50
C-15	100	80	100	80	97	45	60	40	90	45	4	65	40	95	80
C-16	80	70	20	75	70	5	70	100	65	60	0.5	95	45	35	95
C-17	90	88	90	70	96	55	90	60	95	55	30	65	15	100	40
C-18	70	90	100	90	98	80	50	65	100	90	15	100	50	80	95

Table 3.5 Preferences of undergraduate group on all 18 criteria for Question-1

Criteria code	Question-1 (most important and max 100 points and others 1–99 points)														
	UG1	UG2	UG3	UG4	UG5	UG6	UG7	UG8	UG9	UG10	UG11	UG12	UG13	UG14	UG15
C-1	100	99	95	75	78	65	100	100	100	99	85	100	100	100	100
C-2	70	97	50	70	95	95	73	30	80	78	90	50	80	59	70
C-3	79	10	30	40	70	10	20	60	65	85	30	40	55	65	50
C-4	78	30	20	45	68	75	80	50	85	75	70	10	75	58	60
C-5	72	20	0	30	90	45	95	40	60	77	50	30	50	70	80
C-6	77	35	10	55	60	40	55	69	75	74	65	20	70	60	45
C-13	84	84	70	80	85	55	85	89	70	79	75	90	65	85	85
C-14	82	83	60	20	88	20	60	20	55	90	60	0	49	61	50
C-15	65	100	90	90	100	100	75	90	50	100	95	60	97	90	75
C-16	59	95	40	50	75	30	65	66	90	76	40	80	99	62	40
C-17	63	90	80	60	80	85	70	70	10	50	100	70	90	63	55
C-18	98	85	100	100	72	50	50	99	20	80	80	95	92	64	80

Table 3.6 Criteria weights obtained from postgraduate group responses for Question-1

Criteria code	Aggregate rating	Rank order	Weights (%)	Max. and min. weights
C-1	1,350	1	11.5	Maximum weight = 11.5 %
C-2	965	6	8.2	
C-3	897	9	7.7	
C-4	995	5	8.5	Minimum weight = 6.0 %
C-5	901	8	7.7	
C-6	820	11	7.0	
C-13	946	7	8.1	
C-14	706	12	6.0	Difference between max. and min. weight = 5.5 %
C-15	1,021	4	8.7	
C-16	885.5	10	7.6	
C-17	1,039	3	8.9	
C-18	1,173	2	10.0	

Table 3.7 Criteria weights obtained from undergraduate group responses for Question-1

Criteria code	Aggregate rating	Rank order	Weights (%)	Max. and min. weights
C-1	1,396	1	11.5	Maximum weight = 11.5 %
C-2	1,087	5	9.0	
C-3	709	12	5.9	
C-4	879	8	7.3	Minimum weight = 5.9 %
C-5	809	10	6.7	
C-6	810	9	6.7	
C-13	1,181	3	9.7	
C-14	798	11	6.6	Difference between max. and min. weight = 5.6 %
C-15	1,277	2	10.5	
C-16	967	7	8.0	
C-17	1,036	6	8.6	
C-18	1,165	4	9.6	

further consideration. This type of weighting method was giving limited opportunity to the participants to show their preferences on the available watershed management criteria. On the other hand, when the survey participants were asked to select 'least important' criterion and assign 10 points and other important criteria would get multiple of 10 points (Questions 2 and 4). The difference between maximum and minimum weights under this weighting method resulted in a larger difference, which ultimately resulted in a larger range of criteria weights. This weighting method provided opportunity to the survey participants by making them more flexible in deciding which criterion was important compared to the other one.

Table 3.8 Criteria weights obtained from postgraduate group responses for Question-2

Criteria code	Aggregate rating	Rank order	Weights (%)	Max. and min. weights
C-1	8,300	1	28.9	Maximum weight = 28.9 %
C-2	2,895	3	10.1	
C-3	1,380	8	4.8	
C-4	3,465	2	12.1	Minimum weight = 2.8 %
C-5	1,030	10	3.6	
C-6	795	12	2.8	
C-13	2,150	7	7.5	
C-14	910	11	3.2	Difference between max. and min. weight = 26.1 %
C-15	2,268	4	7.9	
C-16	1,210	9	4.2	
C-17	2,160	6	7.5	
C-18	2,189	5	7.6	

Table 3.9 Criteria weights obtained from undergraduate group responses for Question-2

Criteria code	Aggregate rating	Rank order	Weights (%)	Max. and min. weights
C-1	2,455	1	15.4	Maximum weight = 15.4 %
C-2	1,370	6	8.6	
C-3	778	11	4.9	
C-4	926	8	5.8	Minimum weight = 4.3 %
C-5	873	10	5.5	
C-6	682	12	4.3	
C-13	1,490	5	9.3	
C-14	910	9	5.7	Difference between max. and min. weight = 11.1 %
C-15	2,125	2	13.3	
C-16	1,085	7	6.8	
C-17	1,520	4	9.5	
C-18	1,775	3	11.1	

Ranking method was used in Question-5 only. The weights resulted from applying this method in the survey also produced a narrow range of weights. Therefore, the study reveal that the weighting method introduced in Question-2 and Question-4 was resulting better sets of criteria weights and should be preferred in stakeholders' survey on watershed management criteria. However, we should emphasize that this conclusion may further be verified in a larger sample size before it could be implemented in solving a real-world problem related to watershed or water resources management problems.

Table 3.10 Criteria weights obtained from postgraduate group responses for Question-3

Criteria code	Aggregate rating	Rank order	Weights (%)	Max. and min. weights
C-1	1,350	1	11.5	Maximum weight = 11.5 %
C-2	933	8	8.0	
C-3	860	10	7.3	
C-4	1,040	4	8.9	Minimum weight = 5.0 %
C-5	905	9	7.7	
C-6	850	11	7.3	
C-7	1,088	3	9.3	
C-8	965	7	8.2	Difference between max. and min. weight = 6.5 %
C-9	585	12	5.0	
C-10	1,031	5	8.8	
C-11	973	6	8.3	
C-12	1,131	2	9.7	

3.6 Summary on Criteria Weights

This section provides summary of all the three weighting methods which we applied in the survey for assigning weights to the watershed management criteria. We changed position of criteria in the questionnaire to eliminate biasness in the questionnaire on the watershed management criteria (Tables 3.16, 3.17, 3.18 and 3.19).

The survey results reveal that the both groups of respondents' overwhelmly preferred "provide good water quality" criterion for developing watershed management plan. They gave second top priority to "efficient use of resources" criterion with assigning 10.5 % of weights. On the other side, survey participants placed "promote recreational activities" as the least important criterion for developing watershed management plan. The survey respondents thought that the criteria C-3, C-4, and C-5 are equally important for the watershed management plan and assigned 8.9 % weights to these criteria. It is important to mention that these three criteria may further be investigated and replaced with some other most important criteria in future studies so that a better watershed management plan can be developed.

In order to know how many respondents were consistent in their responses while filling out the questionnaire, we evaluated the respondents' responses in a way that Q-1 and Q-2 responses and Q-3 and Q-4 responses for each group of survey participants were combined and their consistency was calculated. We found that undergraduate group was slightly more consistence in the responses with

Table 3.11 Criteria weights obtained from undergraduate group responses for Question-3

Criteria code	Aggregate rating	Rank order	Weights (%)	Max. and min. weights
C-1	1,355	1	11.4	Maximum weight = 11.4 %
C-2	1,035	7	8.7	
C-3	638	11	5.4	
C-4	833	9	7.0	Minimum weight = 4.1 %
C-5	906	8	7.6	
C-6	708	10	6.0	
C-7	1,047	6	8.8	
C-8	1,212	4	10.2	Difference between max. and min. weight = 7.3 %
C-9	488	12	4.1	
C-10	1,093	5	9.2	
C-11	1,272	3	10.7	
C-12	1,306	2	11.0	

Table 3.12 Criteria weights obtained from postgraduate group responses for Question-4

Criteria code	Aggregate rating	Rank order	Weights (%)	Max. and min. weights
C-1	17,860	1	40.2	Maximum weight = 40.2 %
C-2	1,865	4	4.2	
C-3	1,390	9	3.1	
C-4	5,035	3	11.3	Minimum weight = 1.1 %
C-5	1,195	11	2.7	
C-6	1,330	10	3.0	
C-7	1,830	5	4.1	
C-8	1,755	7	3.9	Difference between max. and min. weight = 39.1 %
C-9	495	12	1.1	
C-10	1,795	6	4.0	
C-11	8,230	2	18.5	
C-12	1,670	8	3.8	

consistency level of 77.01 %. The postgraduate group consistency was calculated and found as 75.82 %. Average consistency in both groups' responses was found as 76.42 %, which was an acceptable limit. All this analysis was done to check whether the survey participants understood the questionnaire and were having good knowledge on the watershed management criteria. Table 3.20 reveals individual consistency of each group (postgraduate and undergraduate).

Table 3.13 Criteria weights obtained from undergraduate group responses for Question-4

Criteria code	Aggregate rating	Rank order	Weights (%)	Max. and min. weights
C-1	2,050	3	12.7	Maximum weight = 13.2 %
C-2	1,215	7	7.5	
C-3	625	11	3.9	
C-4	717	10	4.5	Minimum weight = 1.8 %
C-5	865	8	5.4	
C-6	727	9	4.5	
C-7	1,640	6	10.2	
C-8	2,130	1	13.2	Difference between max. and min. weight = 11.4 %
C-9	293	12	1.8	
C-10	1,765	5	11.0	
C-11	1,990	4	12.4	
C-12	2,086	2	13.0	

Table 3.14 Criteria weights obtained from postgraduate group responses for Question-5

Criteria code	Sum of inverse ranks	Rank order	Weights (%)	Max. and min. weights
C-1	237	1	9.4	Maximum weight = 9.4 %
C-2	126	11	5.0	
C-3	124	12	4.9	
C-4	147	7	5.8	
C-5	122	14	4.8	Minimum weight = 2.3 %
C-6	119	16	4.7	
C-7	165	4	6.5	
C-8	121	15	4.8	Difference between max. and min. weight = 7.1 %
C-9	58	18	2.3	
C-10	144	8	5.7	
C-11	156	6	6.2	
C-12	173	3	6.8	
C-13	143	9	5.7	
C-14	82	17	3.2	
C-15	134	10	5.3	
C-16	124	12	4.9	
C-17	160	5	6.3	
C-18	195	2	7.7	

Table 3.15 Criteria weights obtained from undergraduate group responses for Question-5

Criteria code	Sum of inverse ranks	Rank order	Weights (%)	Max. and min. weights
C-1	230	1	8.9	Maximum weight = 8.9 %
C-2	153	9	6.0	
C-3	64	17	2.5	
C-4	95	15	3.7	
C-5	100	14	3.9	Minimum weight = 1.2 %
C-6	73	16	2.8	
C-7	144	11	5.6	
C-8	189	5	7.4	Difference between max. and min. weight = 7.7 %
C-9	31	18	1.2	
C-10	172	8	6.7	
C-11	176	6	6.8	
C-12	221	2	8.6	
C-13	175	7	6.8	
C-14	103	13	4.0	
C-15	197	3	7.7	
C-16	114	12	4.4	
C-17	144	10	5.6	
C-18	190	4	7.4	

3.7 Surveying Popular Databases for the Weighting Methods

We reviewed literature for weighting methods by searching three popular databases including sciencedirect, scopus, and sciverse. In this project our database search was limited to the following criteria:

1. The period was fixed from 2000 to 2012 (13 years)
2. We searched papers in journals only (excluding books, etc.)
3. We searched only four keywords (MCDM, MCDA, MCA, MADM) for search of multi-criteria methods in combination of specific weighting methods
4. Only searched for popular weighting methods (AHP, ranking, pairwise comparison, trade-off, SWING, SMART, Least square method, Delphi, eigenvector).

Total number of retrievals for each weighting method in combination of four other keywords (MCDM, MCDA, MCA, MADM) are listed in Table 3.21. It was found that the 'ranking' was most frequent appearing weighting method in the sciencedirect database with total number of retrievals equal to 3,323 (last column of Table 3.21). AHP (2,060 retrievals) was second most appearing weighting method. Pairwise comparison was the third most appearing weighting method in the

Table 3.16 Criteria weights calculated from preferences of undergraduate group (n = 15)

Criteria code	Q-1 (weights in %)	Q-2 (weights in %)	Q-3 (weights in %)	Q-4 (weights in %)	Q-5 (weights in %)	Average criteria weights (Q1–Q5)	Final rank based on average weights (Q1–Q5)
C-1	11.5	15.4	11.4	12.7	8.9	12.0	1
C-2	9.0	8.6	8.7	7.5	6.0	8.0	10
C-3	5.9	4.9	5.4	3.9	2.5	4.5	17
C-4	7.3	5.8	7.0	4.5	3.7	5.7	14
C-5	6.7	5.5	7.6	5.4	3.9	5.8	13
C-6	6.7	4.3	6.0	4.5	2.8	4.9	16
C-7			8.8	10.2	5.6	8.2	9
C-8			10.2	13.2	7.4	10.3	4
C-9			4.1	1.8	1.2	2.4	18
C-10			9.2	11.0	6.7	9.0	7
C-11			10.7	12.4	6.8	10.0	5
C-12			11.0	13.0	8.6	10.9	2
C-13	9.7	9.3			6.8	8.6	8
C-14	6.6	5.7			4.0	5.4	15
C-15	10.5	13.3			7.7	10.5	3
C-16	8.0	6.8			4.4	6.4	12
C-17	8.6	9.5			5.6	7.9	11
C-18	9.6	11.1			7.4	9.4	6

Table 3.17 Criteria weights calculated from preferences of postgraduate group (n = 15)

Criteria code	Q-1 (weights in %)	Q-2 (weights in %)	Q-3 (weights in %)	Q-4 (weights in %)	Q-5 (weights in %)	Average criteria weights (Q1–Q5)	Final rank based on average weights (Q1–Q5)
C-1	11.5	28.9	11.5	40.2	9.4	20.3	1
C-2	8.2	10.1	8.0	4.2	5.0	7.1	7
C-3	7.7	4.8	7.3	3.1	4.9	5.6	13
C-4	8.5	12.1	8.9	11.3	5.8	9.3	3
C-5	7.7	3.6	7.7	2.7	4.8	5.3	15
C-6	7.0	2.8	7.3	3.0	4.7	4.9	16
C-7			9.3	4.1	6.5	6.6	10
C-8			8.2	3.9	4.8	5.7	12
C-9			5.0	1.1	2.3	2.8	18
C-10			8.8	4.0	5.7	6.2	11
C-11			8.3	18.5	6.2	11.0	2
C-12			9.7	3.8	6.8	6.8	9
C-13	8.1	7.5			5.7	7.1	8
C-14	6.0	3.2			3.2	4.1	17
C-15	8.7	7.9			5.3	7.3	6
C-16	7.6	4.2			4.9	5.6	14
C-17	8.9	7.5			6.3	7.6	5
C-18	10.0	7.6			7.7	8.4	4

Table 3.18 Average criteria weights of both groups (undergraduate and postgraduate, n = 30)

Criteria code	Criteria weights % (undergraduate group) n = 15	Criteria weights % (postgraduate group) n = 15	Average weights % (undergraduate and postgraduate groups) n = 30	Ranking of watershed management criteria
C-1	12.0(1)[a]	20.3(1)	16.2	1
C-2	8.0(10)	7.1(8)	7.6	9
C-3	4.5(17)	5.6(14)	5.1	15
C-4	5.7(14)	9.3(3)	7.5	11
C-5	5.8(13)	5.3(15)	5.6	14
C-6	4.9(16)	4.9(16)	4.9	16
C-7	8.2(9)	6.6(10)	7.4	12
C-8	10.3(4)	5.7(12)	8.0	6
C-9	2.4(18)	2.8(18)	2.6	18
C-10	9.0(7)	6.2(11)	7.6	10
C-11	10.0(5)	11.0(2)	10.5	2
C-12	10.9(2)	6.8(9)	8.9	5
C-13	8.6(8)	7.1(7)	7.9	7
C-14	5.4(15)	4.1(17)	4.8	17
C-15	10.5(3)	7.3(6)	8.9	4
C-16	6.4(12)	5.6(13)	6.0	13
C-17	7.9(11)	7.6(5)	7.8	8
C-18	9.4(6)	8.4(4)	8.9	3

[a] *Note* values shown in brackets in columns 2 and 3 are ranking of criteria

sciencedirect database. On the other hand, 'Simos procedure' has least retrievals (49) which show that not many researchers have used this weighting method in their case studies. We believe that the least number of retrievals for the Simos procedure are for its difficulty in operation and administration in field where majority of stakeholders are generally not very good in mathematics or statistics, etc. It should be noted that the sciencedirect database was searched during November month of 2012 by using online services of PSZ library of Universiti Teknologi Malaysia.

We also searched sciencedirect database for the list of journals that have been publishing papers on different weighting methods. The keywords and the weighting methods searched were kept similar (popular weighting methods in combination of MCDM, MCDA, MCA, and MADM keywords). A total number of 39 journals were retrieved from the sciencedirect database. Table 3.22 shows the list of journals along with the number of retrievals and weight percentage. It can be seen that 'Expert Systems with Application' and 'European Journal of Operational Research were two most frequent appearing journals with total number of retrievals 1,396 (20.49 % weight) and 1,191 (17.48 % weight) respectively.

Table 3.19 Final ranking of watershed management criteria

Final ranking of watershed management criteria	Watershed management criteria	Criteria code	Average criteria weights (%)
1	Provide good water quality	C-1	16.2
2	Efficient use of resources (land and water)	C-11	10.5
3	Improve groundwater and surface water quantity and quality	C-18	8.9
3	Protect woodlands, wetlands and habitats	C-15	8.9
3	Balancing ecological, economic and social interests of watershed	C-12	8.9
6	Protect natural/cultural heritage and forests	C-8	8.0
7	Improve water supply reliabilities to inhabitants	C-13	7.9
8	Safeguarding rivers, lakes and streams	C-17	7.8
10	Provide good habitat for fish and aquatic species	C-2	7.6
11	Protection of sensitive lands (e.g. flood plains and valley lands)	C-10	7.6
12	Provide adequate flood control	C-4	7.5
13	Controlling soil erosion	C-7	7.4
14	Protect life and property from natural hazards	C-16	6.0
15	Provide high quality of life by increasing green space	C-5	5.6
15	Provide economic development incentives	C-3	5.1
17	Provide adequate protection from droughts	C-6	4.9
18	Improve nutrient management	C-14	4.8
19	Promote recreational activities	C-9	2.6

We also analyzed the search results for each weighting method. We found that the application of all weighting methods have increased significantly from 2000 to year 2012. For example, AHP weighting method has total 260 retrievals for 2012 and this figure for year 2000 was just 35 (about 8 times more application in 2012 compared to year 2000). Figure 3.2 shows number of retrievals for AHP weighting method along with four keywords (MCDM, MCDA, MCA, MADM). Figures 3.3, 3.4, 3.5, 3.6, 3.7, 3.8, 3.9 and 3.10 show number of retrievals for other weighting methods for sciencedirect database only.

Table 3.20 Consistency in survey responses

Postgraduate group	Consistency in PG group responses (%)	Undergraduate Group	Consistency in UG group responses (%)	Average consistency of both groups
PG1	91.6	UG1	100.0	76.42 %
PG2	95.8	UG2	75.0	
PG3	75.0	UG3	91.6	
PG4	58.3	UG4	79.2	
PG5	91.6	UG5	75.0	
PG6	83.3	UG6	67.0	
PG7	37.5	UG7	91.6	
PG8	87.5	UG8	67.0	
PG9	79.2	UG9	42.0	
PG10	75.0	UG10	100.0	
PG11	50.0	UG11	87.5	
PG12	79.2	UG12	95.8	
PG13	79.2	UG13	83.0	
PG14	70.8	UG14	63.0	
PG15	83.3	UG15	37.5	
Average consistency	75.82	Average consistency	77.01	

Table 3.21 Weighting methods and number of retrievals for all keywords (MCDM, MCDA, MCA, MADM) (sciencedirect database only)

Rank order (based on last column values)	Weighting method	Number of retrievals for specific weighting method and keyword [searching period = 13 years (2000–2012)]				Column sum
		& MCDM	& MCDA	& MCA	& MADM	
1	Ranking	1,614	680	512	517	3,323
2	AHP	1,192	364	149	355	2,060
3	Pairwise comparison	680	264	230	157	1,331
4	Trade off	453	234	285	91	1,063
5	SMART and SMARTER	244	131	363	67	805
6	SWING	60	65	160	10	295
7	Least square method	88	25	126	27	266
8	Delphi method	124	32	26	41	223
9	Eigenvector method	103	26	10	42	181
10	Simple multi-attribute rating technique	33	20	7	10	70
11	SIMOS Procedure	19	16	3	11	49
	Row sum	4,610	1,857	1,871	1,328	9,666

Table 3.22 Journal versus number of search retrievals (sciencedirect database only)

Rank order	Journal	Total retrievals (period 2000–2012)	Weight (%)
1	Expert systems with applications	1,396	20.49
2	European journal of operational research	1,191	17.48
3	Mathematical and computer modelling	319	4.68
4	Omega	266	3.90
5	Journal of environmental radioactivity	261	3.83
6	Applied soft computing	250	3.67
7	Information sciences	242	3.55
8	Computers and operations research	235	3.45
9	Decision support systems	234	3.44
10	Energy policy	223	3.27
11	International journal of production economics	187	2.75
12	Computers and industrial engineering	179	2.63
13	Ecological economics	171	2.51
14	Knowledge-based systems	166	2.44
15	Applied mathematical modelling	163	2.39
16	Materials and design	160	2.35
17	Environmental modelling and software	131	1.92
18	Journal of cleaner production	88	1.29
19	Fuzzy sets and systems	87	1.28
20	Renewable and sustainable energy reviews	82	1.20
21	Science of the total environment	79	1.16
22	Automation in construction	69	1.01
23	Applied mathematics and computation	61	0.90
24	Fuel and energy abstracts	58	0.85
25	Energy	57	0.84
26	Land use policy	55	0.81
27	Nuclear instruments and methods in physics research	55	0.81
28	Waste management	48	0.70
29	Forest ecology and management	44	0.65
30	Ecological indicators	37	0.54
31	Environmental impact assessment review	35	0.51
32	Brain research	33	0.48
33	Computer communications	33	0.48
34	Procedia—social and behavioral sciences	30	0.44
35		29	0.43

(continued)

Table 3.22 (continued)

Rank order	Journal	Total retrievals (period 2000–2012)	Weight (%)
	International journal of project management		
36	Value in health	27	0.40
37	Building and environment	15	0.22
38	Biological conservation	12	0.18
39	Ecological modelling	4	0.06

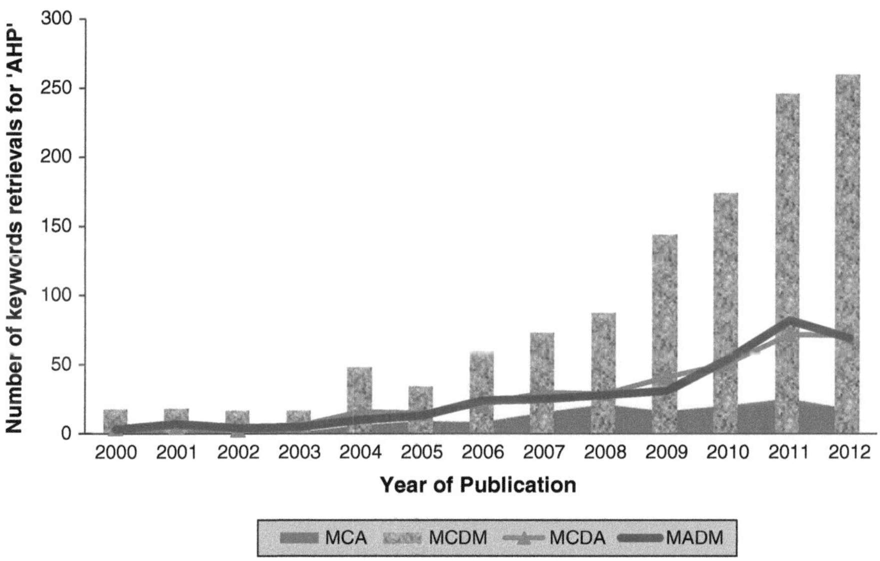

Fig. 3.2 Number of keywords (MCDM, MCDA, MCA, MADM) retrievals in combination of "AHP" (sciencedirect database only)

We also analyzed the retrieval data for determining cumulative appearance of different weighting methods in sciencedirect database. We found that there is significant increase in trend for using MCDM short term for multi-criteria methods. This analysis also shows that almost all weighting methods are appearing more and more in number since 2000 which indicates that many researchers have been using weighting methods in their multi-criteria decision analysis problems. Cumulative number of retrievals for different keywords and weighting methods are shown in Figs. 3.11, 3.12, 3.13, 3.14, 3.15, 3.16, 3.17, 3.18 and 3.19.

The number of retrievals of weighting methods and keywords for scopus and sciverse databases are shown in Appendix C.

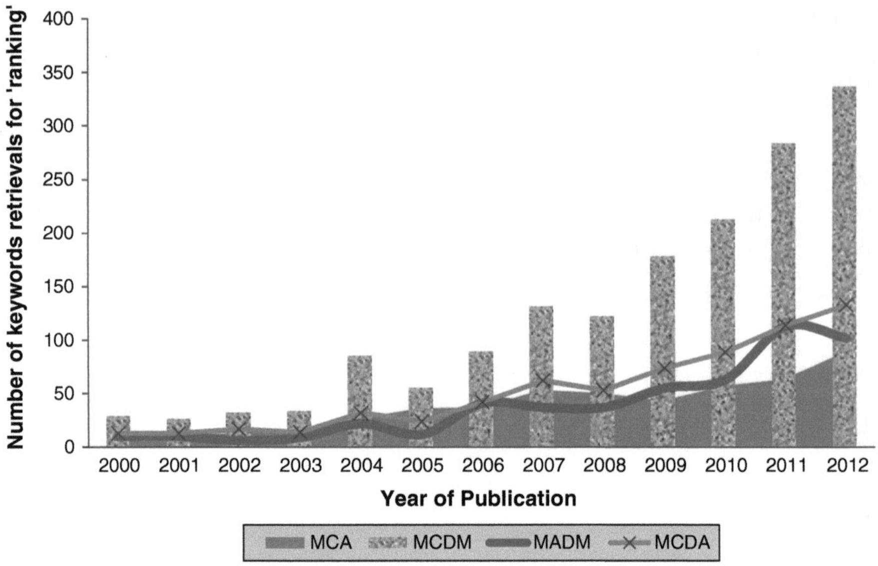

Fig. 3.3 Number of keywords (MCDM, MCDA, MCA, MADM) retrievals in combination of "Ranking" (sciencedirect database only)

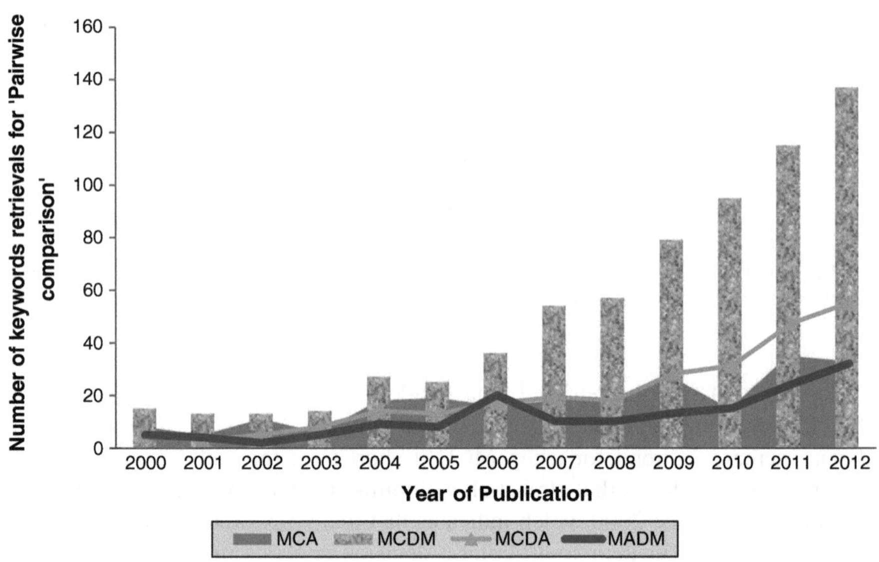

Fig. 3.4 Number of keywords (MCDM, MCDA, MCA, MADM) retrievals in combination of "Pairwise comparison" (sciencedirect database only)

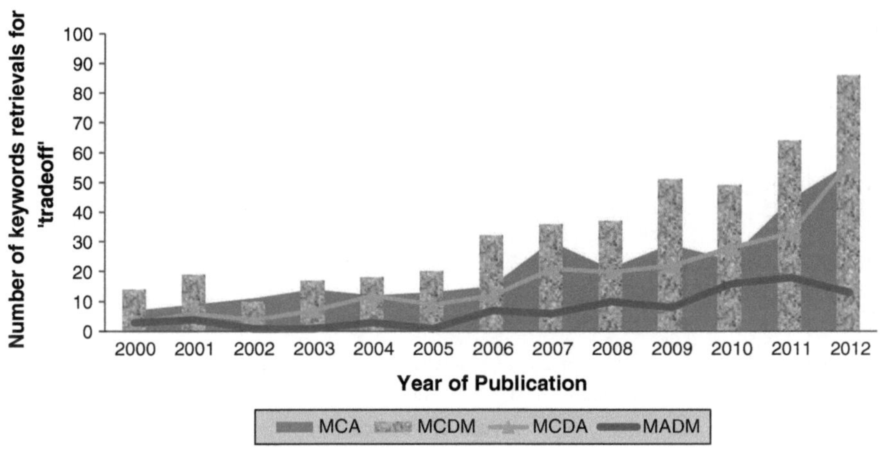

Fig. 3.5 Number of keywords (MCDM, MCDA, MCA, MADM) retrievals in combination of "Tarde-off" (sciencedirect database only)

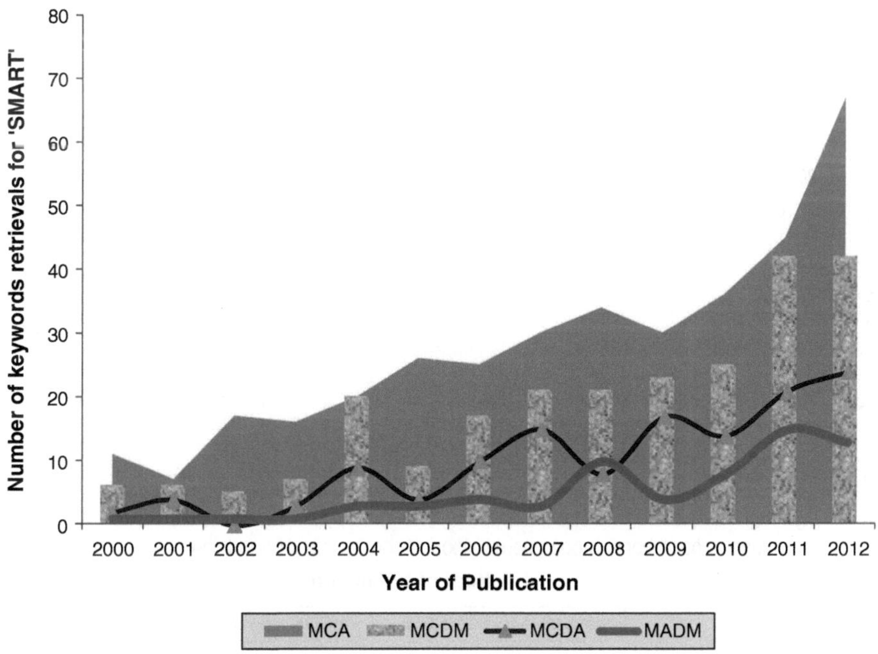

Fig. 3.6 Number of keywords (MCDM, MCDA, MCA, MADM) retrievals in combination of "SMART" (sciencedirect database only)

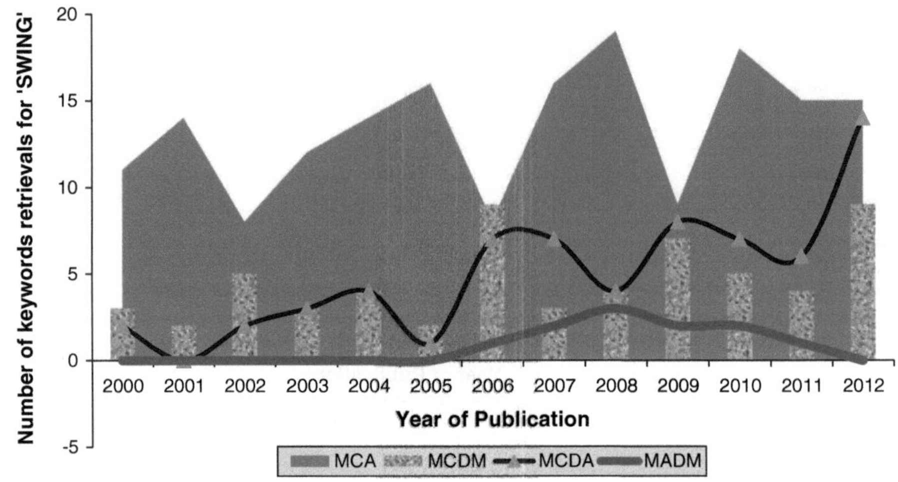

Fig. 3.7 Number of keywords (MCDM, MCDA, MCA, MADM) retrievals in combination of "SWING" (sciencedirect database only)

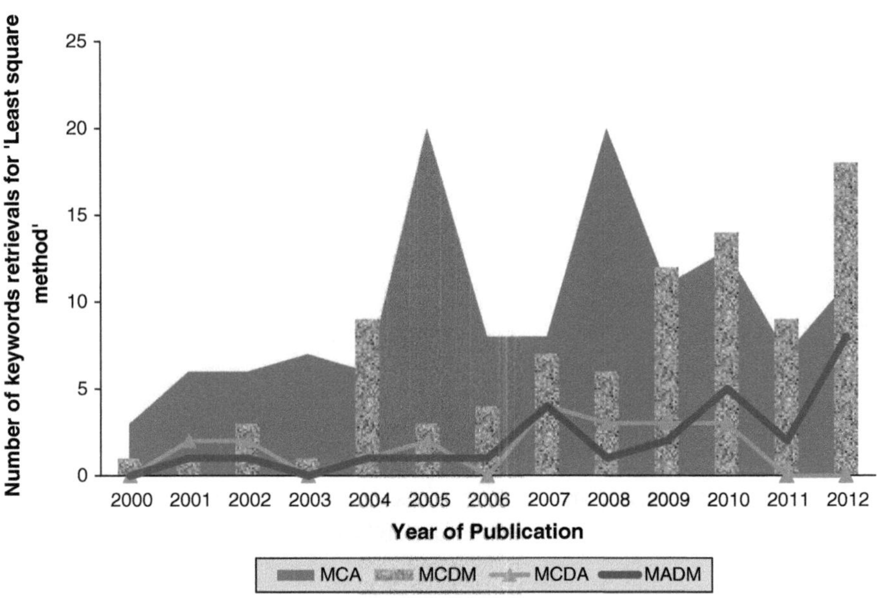

Fig. 3.8 Number of keywords (MCDM, MCDA, MCA, MADM) retrievals in combination of "Least square method" (sciencedirect database only)

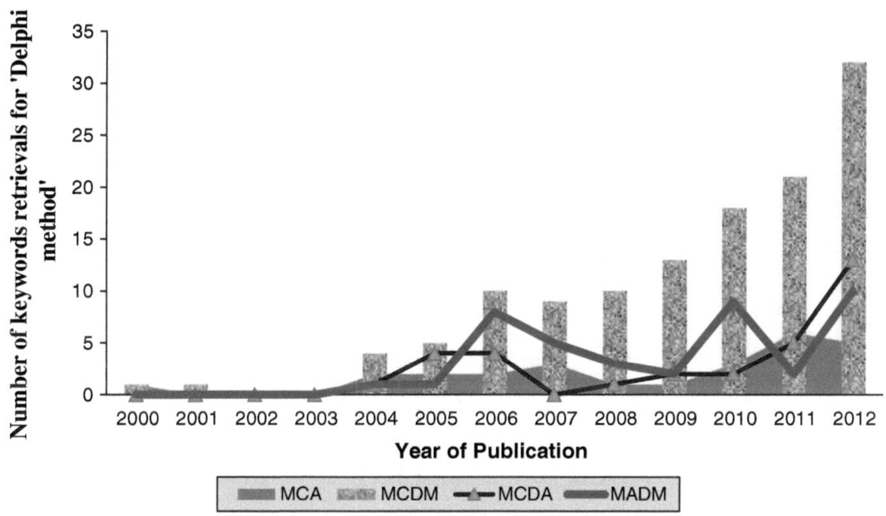

Fig. 3.9 Number of keywords (MCDM, MCDA, MCA, MADM) retrievals in combination of "Delphi method" (sciencedirect database only)

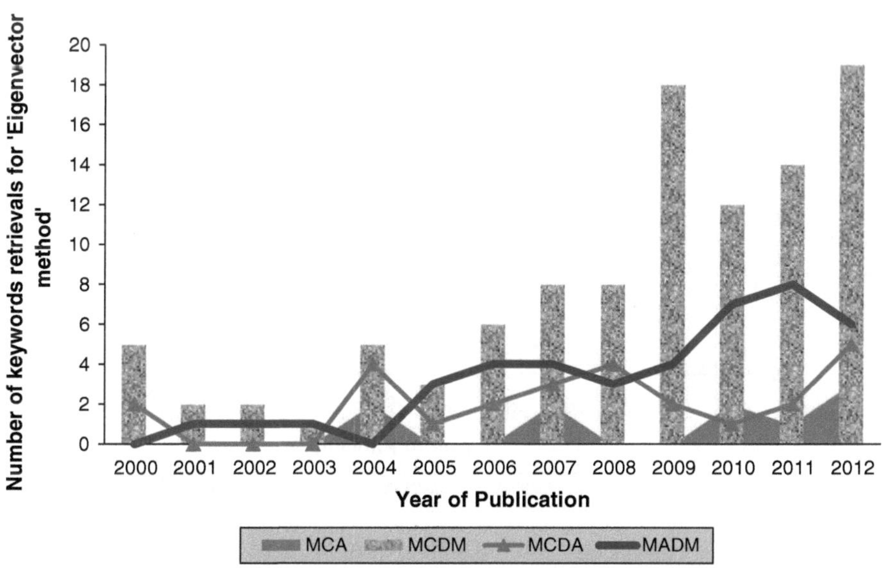

Fig. 3.10 Number of keywords (MCDM, MCDA, MCA, MADM) retrievals in combination of "Eigenvector method" (sciencedirect database only)

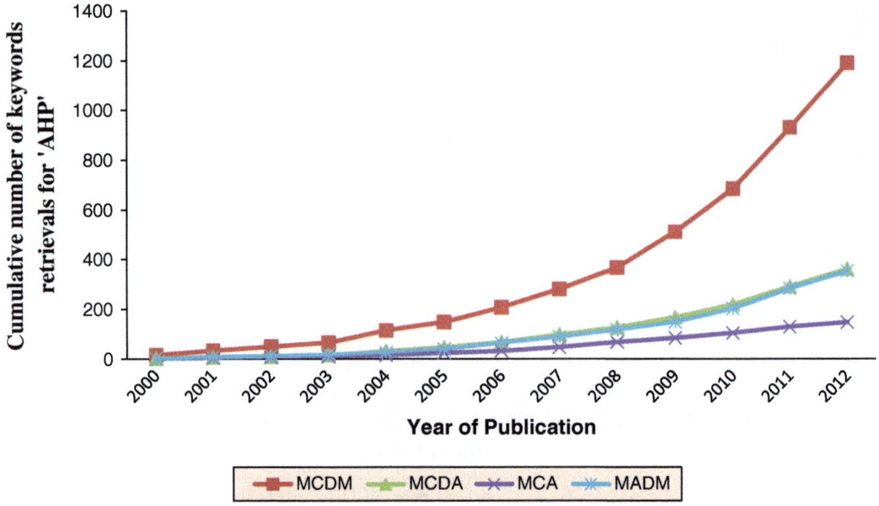

Fig. 3.11 Cumulative plot of keywords (MCDM, MCDA, MCA, MADM) in combination with "AHP" keyword (sciencedirect database only)

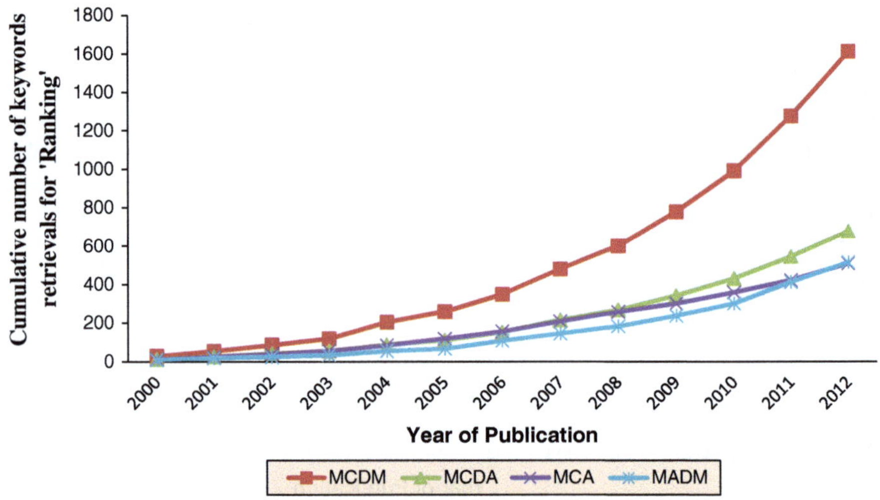

Fig. 3.12 Cumulative plot of keywords (MCDM, MCDA, MCA, MADM) in combination with "Ranking" keyword (sciencedirect database only)

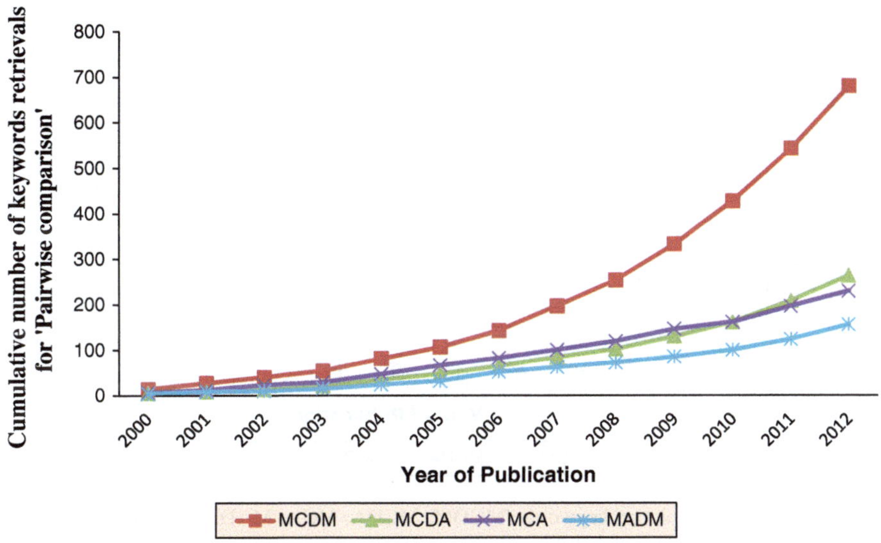

Fig. 3.13 Cumulative plot of keywords (MCDM, MCDA, MCA, MADM) in combination with "Pairwise comparision" keyword (sciencedirect database only)

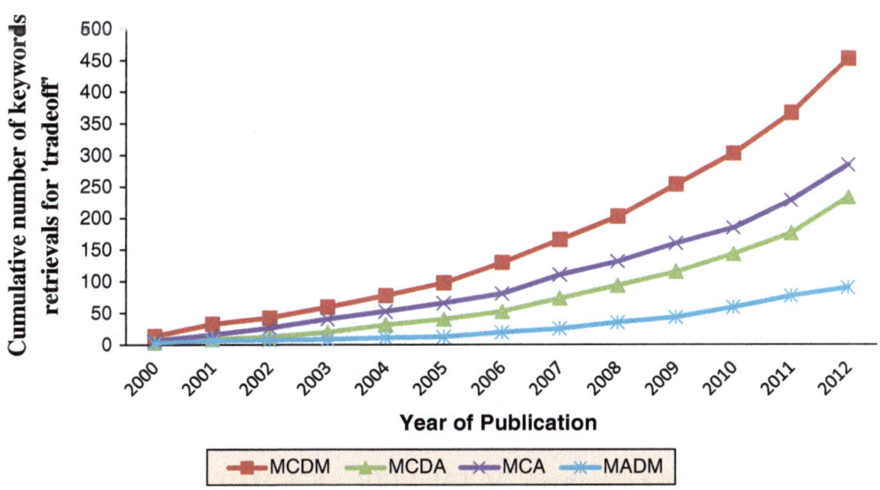

Fig. 3.14 Cumulative plot of keywords (MCDM, MCDA, MCA, MADM) in combination with "Trade-off" keyword (sciencedirect database only)

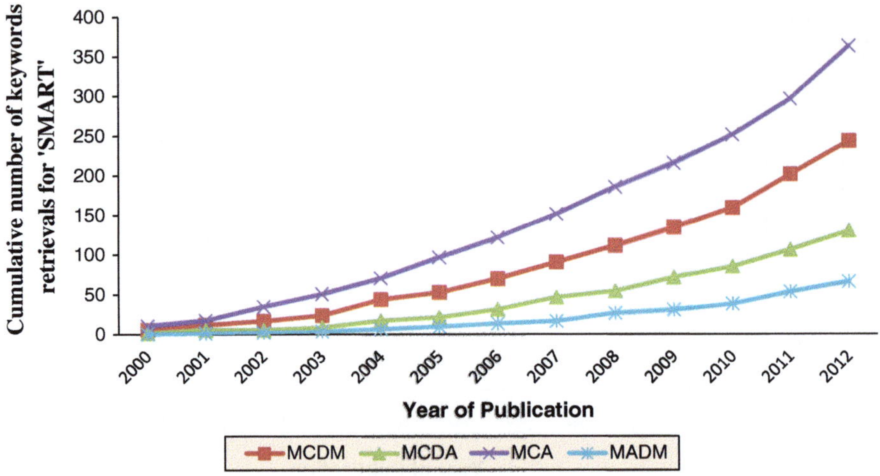

Fig. 3.15 Cumulative plot of keywords (MCDM, MCDA, MCA, MADM) in combination with "SMART" keyword (sciencedirect database only)

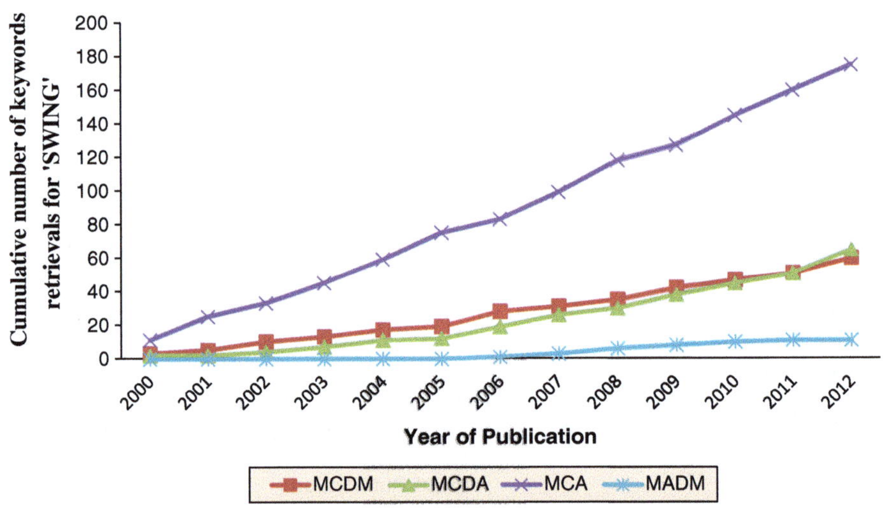

Fig. 3.16 Cumulative plot of keywords (MCDM, MCDA, MCA, MADM) in combination with "SWING" keyword (sciencedirect database only)

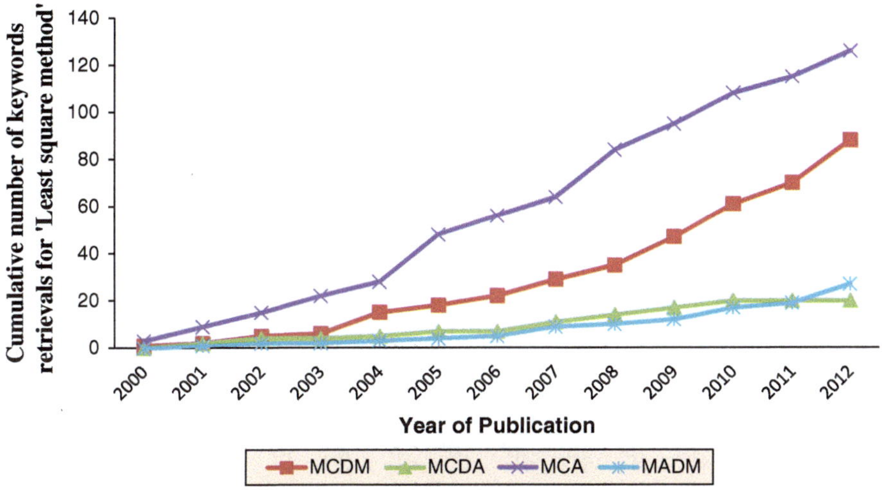

Fig. 3.17 Cumulative plot of keywords (MCDM, MCDA, MCA, MADM) in combination with "Least square method" keyword (sciencedirect database only)

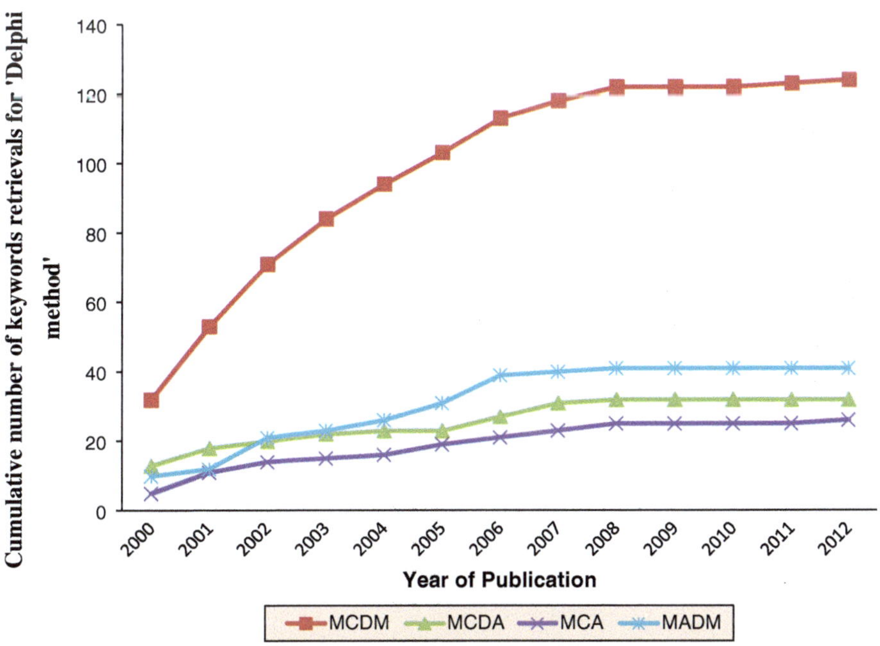

Fig. 3.18 Cumulative plot of keywords (MCDM, MCDA, MCA, MADM) in combination with "Delphi" keyword (sciencedirect database only)

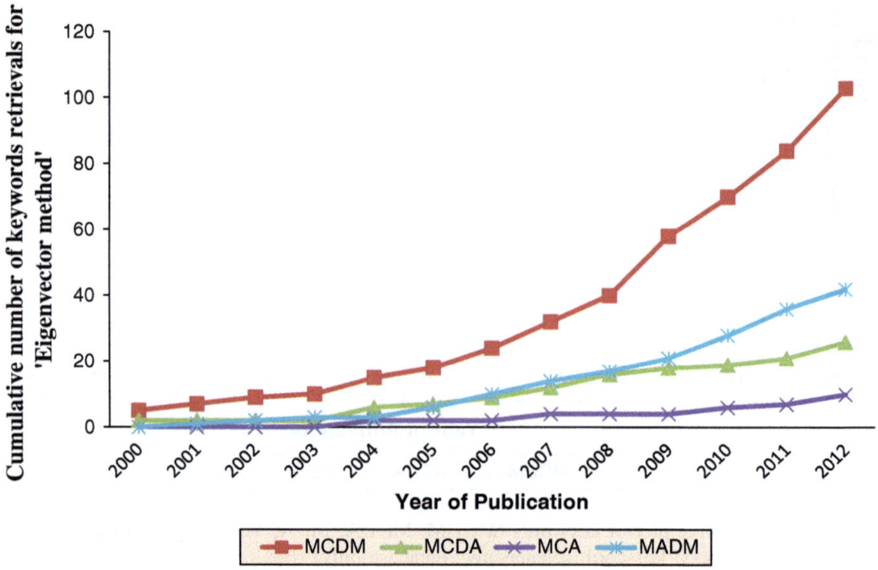

Fig. 3.19 Cumulative plot of keywords (MCDM, MCDA, MCA, MADM) in combination with "Eigenvector method" keyword (sciencedirect database only)

Chapter 4
Conclusions and Recommendations

4.1 Chapter Summary

This chapter presents main findings of the research carried out under this project. The conclusions drawn from the survey data analysis and data mining using three main scholar databases are also presented in this chapter. Finally, recommendations for further study and new research directions in this field of research are given at the end of the chapter. Here, we suggest that these conclusions are based on data analysis of only 30 respondents on watershed management criteria and data mining by using only three scholar databases namely sciencedirect, scopus, and sciverse using some specific keywords for different popular weighting methods and common abbreviations generally used for representing multi-criteria decision-making methods.

4.2 Conclusions

'Provide good water quality' was found the most important watershed management criterion to the both groups (postgraduate and undergraduate students) of survey as the aggregate weight assigned to this criterion was calculated as 16.2 %. The survey data analysis on stakeholders' preferences on watershed management criteria found that 'efficient use of land and water resources' was the second most important criterion to the survey participants with assigned weight equal to 10.5 %. Three criteria namely 'improve groundwater and surface water quantity and quality', 'protect woodlands, wetlands and habitats', and 'balancing ecological, economic and social interests of watershed' were given equal importance by the survey participants with calculated weight of 8.9 %. As expected, 'promote recreational activities' was found the least important criterion of watershed management with given weight merely 2.6 %. We must emphasize that this type of findings are new in

N.H. Zardari et al., *Weighting Methods and their Effects on Multi-Criteria Decision Making Model Outcomes in Water Resources Management*,
SpringerBriefs in Water Science and Technology, DOI 10.1007/978-3-319-12586-2_4

its nature as previous research was lacking of such priority weights of watershed management criteria.

We found that different weighting methods applied to the same problem produced variety of weights attached to watershed management criteria. These different sets of weights may have significant effects on the final results of the problem chosen for solving any problem in the fields of watershed management and water resource planning and management. Some weighting methods produced narrow weights as these weighting methods were rigid in nature where respondents may not had much flexibility to show their preferences on given watershed management criteria. These weighting methods include point allocation and ranking methods. On the other hand, some weighting methods produced a reasonable range of criteria weights. These weighting methods are minimum allocation of 10 points to the least important criterion and multiple of this to other criteria which were thought to be more important than the criterion assigned 10 points.

We conclude that the choice of weighting method for solving a real world watershed management problem is crucial and much attention of the researchers and decision makers is required before proceeding with next steps in the problem solving framework. This conclusion is based on survey findings of only 30 participants on 18 watershed management criteria. We emphasize that these weights are not stable and may change if number of criteria are increased or decreased from 18 watershed management criteria and if the number of participants are increased or decreased from 30. These weights may also change if preferences on watershed management criteria are obtained from more expert respondents compared to postgraduate and undergraduate students which took part in our survey under the current project.

We also conclude that the weighting method introduced in Question-2 and Question-4 of the survey questionnaire resulted better sets of criteria weights and should be preferred in future studies on determining stakeholders' priorities on watershed management criteria. However, we should emphasize that this conclusion may further be verified in a larger sample size before it could be implemented in solving a real-world problem related to watershed or water resources management problems.

The survey data analysis for individual groups of postgraduate and undergraduate students confirmed our hypothesis that the postgraduate students may have more understanding on the watershed management criteria compared to undergraduate students who may not have more knowledge on watershed phenomenon even though we explained both groups on the watershed management criteria and the purpose of survey. The postgraduate group produced a wider range of weights on watershed management criteria compared to undergraduate students who were found more conservative or stringent on allocating points to different criteria in the survey. These conclusions can be confirmed from the survey results presented in Table 3.17 given in the body of the thesis (see Chap. 3 for Table 3.17 and other similar tables).

However, the survey data analysis found that the undergraduate group was slightly more consistent in their survey responses with consistency level of 77.01 %

compared to the postgraduate group consistency which was found slightly lower (75.82 %). Average consistency in both groups' responses was found as 76.42 %, which was an acceptable limit. We should remind here that this type of consistency check on survey data was part of sensitivity analysis of the survey data to confirm how both groups were consistent in their responses in answering questions of the survey questionnaire.

In order to see which weighting methods were very popular and which weighting methods were not very popular and applied in limited number of studies in solving research or real world watershed, water resource, and other multi-criteria decision problems, we reviewed literature for weighting methods by searching three popular databases including sciencedirect, scopus, and sciverse. Our data mining was limited to that following keywords and search period:

1. The period was fixed from 2000 to 2012 (13 years)
2. We searched papers in journals only (excluding books, etc.)
3. We searched only four keywords (MCDM, MCDA, MCA, MADM) for search of multi-criteria methods in combination of specific weighting methods
4. Only searched for popular weighting methods (AHP, ranking, pairwise comparison, trade-off, SWING, SMART, Least square method, Delphi, eigenvector)

Data mining found that the 'ranking' was most frequent appearing weighting method in all three databases (sciencedirect, scopus, and sciverse). Review of weighting methods by using only sciencedirect shows that the 'ranking' was most frequently applied weighting method with total number of retrievals equal to 3,323 (last column of Table 3.21). AHP (2,060 retrievals) was second most appearing weighting method in sciencedirect database. Pairwise comparison was the third most appearing weighting method in the sciencedirect database. On the other hand, 'Simos procedure' has least retrievals (49) which shows that not many researchers have used this weighting methods in their case studies. We believe that the least number of retrievals for the Simos procedure are for its difficulty in operation and administration in field where majority of stakeholders are generally not very good in mathematics or statistics, etc. It should be noted that all three scholar databases were searched during November month of 2012 by using online services of PSZ library of Universiti Teknologi Malaysia.

To check which journal publishing articles on multi-criteria decision-making methods and weighting methods, we searched all three databases. The keywords and the weighting methods searched were kept similar (popular weighting methods in combination of MCDM, MCDA, MCA, and MADM keywords). A total number of 39 journals were retrieved from the sciencedirect database. Table 3.22 shows the list of journals along with the number of retrievals and weight percentage. It can be seen that 'Expert Systems with Application' and 'European Journal of Operational Research' were two most frequent appearing journals with total number of retrievals 1,396 (20.49 % weight) and 1,191 (17.48 % weight) respectively. Appendix C shows data mining outputs for scopus and sciverse databases for famous weighting methods and popular journals which were publishing articles on weighting methods.

We also analyzed the search results for each weighting method. We found that the application of all weighting methods have increased significantly from 2000 to year 2012. For example, AHP weighting method has total 260 retrievals for 2012 and this figure for year 2000 was just 35 (about 8 times more application in 2012 compared to year 2000). This trend in increase application of weighting methods since year 2000 was also observed while mining data from scopus and sciverse databases. This clearly shows that many researchers around the world are frequently using weighting methods and multi-criteria decision making methods in their studies. We also predict that this trend in increase usage of multi-criteria decision making methods and weighting methods will continue in future as even simple decisions in water resources could be made by using some decision tools to convince stakeholders that decisions are being made by following a recognized procedure. This prediction is not merely based on the findings of data mining from three databases in this study, but this it is also based on the complications arising in making water resource decisions when competition for getting water increasing many-fold.

4.3 Recommendations

As mentioned in earlier sections of this chapter that data mining was performed only for 13 years (2000–2011), this period can further be extended for at least 30 years to get more stable results on application of weighting methods and their priority ranks. Because of the time limitation to complete this study, we only focused on few popular weighting methods rather than searching for an exhaustive list of methods covering all weighting methods and more and more keywords for multi-criteria decision-making methods. We again emphasize that the current study provides only partial guideline on how to choose weighting method; rather than giving a complete guidance on all factors associated to the selection of weighting methods. Therefore, we recommend few directions for future research in this field. These recommendations are:

1. Bigger period for searching weighting methods could be selected in future studies. We suggest that 30 years is initially a reasonable period for data mining.
2. We used only three databases which were available to us through Universiti Teknologi Universiti library. However, in future, more databases could be used. For example, ISI Web of Science and ISI Web of Knowledge, etc.
3. Watershed management criteria may be revisited to check whether any other important criteria were not missing. However, we believe that 18 criteria selected for this study are reasonably a good number to decide which watershed needs more attention for funding or some rehabilitation or management measures and activities.
4. The number of survey sample was very less than the minimum number of samples which may need to be considered for determining more stable weights

of watershed management criteria. Therefore, in future studies, we recommend that at least 200 samples may be presented the list of watershed management criteria to get priority weights.

5. In this study, our samples were undergraduate and postgraduate students. Their knowledge on real world problems related to watershed may not be enough to judge anything with high certainty in this field of research. We, therefore, suggest that future studies may ask more experienced people to participate in the survey and show their preferences on watershed management criteria. We believe that the priority ranks obtained for watershed management criteria from such experienced people will be stable and will give results very close to the reality.
6. We only applied three methods (ranking, point allocation, and ratio methods) in this study. We recommend that future studies may review at least ten popular weighting methods including pairwise weighting method and test their outputs in field from a greater number of samples as it may produce a better ranking of weighting methods. This practice will also help decision-makers to choose the best weighting method for their decision-making process.

[illegible] Therefore, in Figure [illegible]

[illegible] samples [illegible] the [illegible] of [illegible]

[illegible]

[illegible]

Appendix A

Survey raw data on postgraduate and undergraduate students' preferences on watershed management criteria for all five survey questions.

Postgraduate Students

(See Tables A.1, A.2, A.3, A.4, A.5, A.6, A.7, A.8, A.9, A.10, A.11, A.12, A.13, A.14 and A.15).

Table A.1 Responses of postgraduate (PG) students on the presented watershed management criteria

PG1					
Criteria	Q1	Q2	Q3	Q4	Q5
C1	90	120	80	120	4
C2	80	100	70	80	12
C3	70	50	50	60	15
C4	50	50	50	50	17
C5	50	30	60	50	16
C6	50	10	50	10	18
C7			80	100	11
C8			90	80	13
C9			70	60	14
C10			90	150	6
C11			100	150	2
C12			90	150	7
C13	60	90			8
C14	60	90			9
C15	100	150			1
C16	80	120			10
C17	90	120			3
C18	70	100			5

N.H. Zardari et al., *Weighting Methods and their Effects on Multi-Criteria Decision Making Model Outcomes in Water Resources Management*,
SpringerBriefs in Water Science and Technology, DOI 10.1007/978-3-319-12586-2

Table A.2 Responses of postgraduate (PG) students on the presented watershed management criteria

PG2					
Criteria	Q1	Q2	Q3	Q4	Q5
C1	100	500	100	500	1
C2	95	400	95	400	2
C3	40	10	40	30	17
C4	70	120	60	70	16
C5	60	80	65	100	13
C6	50	30	50	50	14
C7			88	200	4
C8			90	280	3
C9			30	10	18
C10			85	180	7
C11			80	150	6
C12			89	250	5
C13	75	100			11
C14	55	60			15
C15	80	200			9
C16	70	150			12
C17	88	250			10
C18	90	300			8

Table A.3 Responses of postgraduate (PG) students on the presented watershed management criteria

PG3					
Criteria	Q1	Q2	Q3	Q4	Q5
C1	100	1980	95	1280	1
C2	80	960	98	320	5
C3	80	240	85	320	9
C4	70	960	80	80	14
C5	65	60	70	120	11
C6	10	10	10	10	18
C7			30	40	15
C8			100	640	2
C9			70	50	12
C10			90	640	7
C11			20	20	16
C12			90	160	6
C13	95	960			10
C14	70	120			13
C15	100	960			3
C16	20	20			17
C17	90	480			8
C18	100	480			4

Table A.4 Responses of postgraduate (PG) students on the presented watershed management criteria

PG4					
Criteria	Q1	Q2	Q3	Q4	Q5
C1	100	1280	100	100	1
C2	85	640	80	40	6
C3	50	10	65	20	17
C4	75	80	90	70	4
C5	60	40	85	30	16
C6	70	80	80	70	5
C7			90	90	2
C8			75	50	7
C9			60	10	18
C10			95	50	8
C11			70	60	9
C12			93	80	3
C13	65	20			15
C14	60	160			14
C15	80	320			10
C16	75	320			12
C17	70	640			11
C18	90	160			13

Table A.5 Responses of postgraduate (PG) students on the presented watershed management criteria

PG5					
Criteria	Q1	Q2	Q3	Q4	Q5
C1	100	250	100	250	1
C2	85	110	80	130	9
C3	30	50	60	60	14
C4	50	30	55	50	16
C5	99	210	90	220	3
C6	60	10	50	20	17
C7			95	190	2
C8			85	160	5
C9			45	10	18
C10			75	110	10
C11			65	70	13
C12			70	80	11
C13	80	90			12
C14	90	130			8
C15	97	150			4
C16	70	70			15
C17	96	170			6
C18	98	190			7

Table A.6 Responses of postgraduate (PG) students on the presented watershed management criteria

PG6					
Criteria	Q1	Q2	Q3	Q4	Q5
C1	100	100	100	100	1
C2	60	65	60	60	8
C3	90	90	75	75	4
C4	50	30	50	65	9
C5	70	70	70	70	5
C6	20	15	30	40	14
C7			90	95	2
C8			80	80	7
C9			0	10	18
C10			85	90	3
C11			10	25	10
C12			40	30	12
C13	5	25			15
C14	15	20			16
C15	45	50			11
C16	5	10			17
C17	55	40			13
C18	80	60			6

Table A.7 Responses of postgraduate (PG) students on the presented watershed management criteria

PG7					
Criteria	Q1	Q2	Q3	Q4	Q5
C1	100	100	100	120	1
C2	65	80	20	10	18
C3	80	70	80	80	3
C4	75	60	55	20	17
C5	92	70	60	60	5
C6	85	75	50	30	16
C7			95	100	6
C8			75	40	15
C9			35	50	14
C10			96	90	7
C11			98	70	8
C12			89	110	9
C13	76	80			11
C14	45	85			13
C15	60	88			10
C16	70	60			12
C17	90	70			4
C18	50	84			2

Table A.8 Responses of postgraduate (PG) students on the presented watershed management criteria

PG8					
Criteria	Q1	Q2	Q3	Q4	Q5
C1	45	30	20	10	14
C2	28	10	25	15	16
C3	80	35	40	30	6
C4	20	15	30	20	13
C5	70	70	80	70	3
C6	30	10	35	25	15
C7			50	40	8
C8			60	50	12
C9			55	45	11
C10			45	35	17
C11			65	55	10
C12			100	90	2
C13	50	60			4
C14	35	20			7
C15	40	25			18
C16	100	90			1
C17	60	40			5
C18	65	45			9

Table A.9 Responses of postgraduate (PG) students on the presented watershed management criteria

PG9					
Criteria	Q1	Q2	Q3	Q4	Q5
C1	80	110	100	100	1
C2	55	40	25	30	14
C3	50	20	10	10	17
C4	70	50	60	70	5
C5	85	80	90	100	8
C6	60	60	80	90	6
C7			70	80	4
C8			15	15	16
C9			20	20	15
C10			50	60	9
C11			40	50	12
C12			30	40	13
C13	75	90			3
C14	45	10			18
C15	90	70			11
C16	65	30			10
C17	95	100			7
C18	100	120			2

Table A.10 Responses of postgraduate (PG) students on the presented watershed management criteria

PG10					
Criteria	Q1	Q2	Q3	Q4	Q5
C1	100	100	100	100	1
C2	40	30	65	60	16
C3	50	45	70	50	8
C4	70	70	80	70	4
C5	30	10	30	10	18
C6	65	60	70	65	5
C7			60	25	12
C8			55	30	13
C9			50	40	14
C10			40	20	15
C11			75	80	11
C12			90	90	3
C13	80	50			7
C14	45	15			17
C15	45	20			10
C16	60	40			6
C17	55	50			9
C18	90	90			2

Table A.11 Responses of postgraduate (PG) students on the presented watershed management criteria

PG11					
Criteria	Q1	Q2	Q3	Q4	Q5
C1	100	3000	100	14400	1
C2	2	50	15	30	8
C3	12	100	5	15	14
C4	60	1000	70	3600	3
C5	20	60	10	10	16
C6	70	30	40	360	6
C7			60	400	4
C8			50	180	18
C9			35	60	15
C10			80	200	7
C11			85	7200	2
C12			20	90	17
C13	10	270			5
C14	1	90			11
C15	4	15			12
C16	0.5	10			13
C17	30	20			9
C18	15	180			10

Table A.12 Responses of postgraduate (PG) students on the presented watershed management criteria

PG12					
Criteria	Q1	Q2	Q3	Q4	Q5
C1	70	70	80	40	10
C2	40	20	40	20	16
C3	55	30	80	40	13
C4	80	60	85	80	8
C5	40	10	40	10	18
C6	65	40	90	90	3
C7			70	50	12
C8			70	50	14
C9			50	30	15
C10			70	50	5
C11			60	70	4
C12			100	100	1
C13	85	40			9
C14	40	20			17
C15	65	50			11
C16	95	60			6
C17	65	50			7
C18	100	80			2

Table A.13 Responses of postgraduate (PG) students on the presented watershed management criteria

PG13					
Criteria	Q1	Q2	Q3	Q4	Q5
C1	85	360	85	400	3
C2	70	120	75	350	5
C3	90	540	90	450	2
C4	100	720	100	500	1
C5	65	60	70	300	6
C6	80	180	65	250	4
C7			45	160	8
C8			30	40	14
C9			50	80	9
C10			40	20	17
C11			35	10	13
C12			60	180	7
C13	60	40			10
C14	20	30			18
C15	40	30			16
C16	45	20			12
C17	15	10			15
C18	50	20			11

Table A.14 Responses of postgraduate (PG) students on the presented watershed management criteria

PG14					
Criteria	Q1	Q2	Q3	Q4	Q5
C1	80	100	90	140	7
C2	90	80	90	130	5
C3	50	50	40	60	11
C4	60	30	80	100	9
C5	10	10	25	20	17
C6	30	15	60	40	15
C7			85	110	8
C8			20	30	16
C9			10	10	18
C10			70	80	13
C11			100	150	1
C12			95	145	2
C13	40	55			12
C14	75	40			10
C15	95	90			4
C16	35	20			14
C17	100	110			3
C18	80	100			6

Table A.15 Responses of postgraduate (PG) students on the presented watershed management criteria

PG15					
Criteria	Q1	Q2	Q3	Q4	Q5
C1	100	200	100	200	1
C2	90	190	95	190	6
C3	70	40	70	90	11
C4	95	190	95	190	2
C5	85	170	60	25	8
C6	75	170	90	180	5
C7			80	150	7
C8			70	30	9
C9			5	10	18
C10			20	20	16
C11			70	70	12
C12			75	75	14
C13	90	180			10
C14	50	20			17
C15	80	50			13
C16	95	190			4
C17	40	10			15
C18	95	180			3

Undergraduate Students

(See Tables A.16, A.17, A.18, A.19, A.20, A.21, A.22, A.23, A.24, A.25, A.26, A.27, A.28, A.29 and A.30).

Table A.16 Responses of undergraduate (UG) students on the presented watershed management criteria

UG1					
Criteria	Q1	Q2	Q3	Q4	Q5
C-1	100	200	100	200	1
C-2	70	80	53	10	15
C-3	79	120	68	50	11
C-4	78	111	64	16	13
C-5	72	90	70	32	12
C-6	77	110	63	15	14
C-7			99	180	3
C-8			98	160	4
C-9			73	78	10
C-10			84	90	8
C-11			92	120	5
C-12			80	81	9
C-13	84	150			6
C-14	82	145			7
C-15	65	20			16
C-16	59	10			18
C-17	63	15			17
C-18	98	180			2

Table A.17 Responses of undergraduate (UG) students on the presented watershed management criteria

UG2					
Criteria	Q1	Q2	Q3	Q4	Q5
C-1	99	170	91	150	10
C-2	97	160	90	160	7
C-3	10	10	10	10	18
C-4	30	50	30	20	16
C-5	20	40	93	100	13
C-6	35	60	40	30	17
C-7			50	50	14
C-8			97	170	5
C-9			80	20	15
C-10			98	180	8
C-11			99	190	2
C-12			100	200	1
C-13	84	100			12
C-14	83	120			11
C-15	100	200			3
C-16	95	180			6
C-17	90	190			4
C-18	85	150			9

Table A.18 Responses of undergraduate (UG) students on the presented watershed management criteria

UG3					
Criteria	Q1	Q2	Q3	Q4	Q5
C-1	95	110	95	140	3
C-2	50	60	40	60	11
C-3	30	40	20	40	16
C-4	20	20	10	20	17
C-5	0	30	30	30	15
C-6	10	10	0	10	18
C-7			100	150	1
C-8			70	70	12
C-9			60	65	14
C-10			90	100	8
C-11			80	130	7
C-12			50	95	9
C-13	70	80			6
C-14	60	70			10
C-15	90	100			4
C-16	40	50			13
C-17	80	90			5
C-18	100	120			2

Table A.19 Responses of undergraduate (UG) students on the presented watershed management criteria

UG4					
Criteria	Q1	Q2	Q3	Q4	Q5
C-1	75	120	90	120	4
C-2	70	110	80	110	5
C-3	40	40	25	20	17
C-4	45	60	45	40	15
C-5	30	20	30	60	16
C-6	55	50	35	50	13
C-7			40	30	14
C-8			50	70	11
C-9			20	10	18
C-10			60	90	7
C-11			70	100	12
C-12			100	150	2
C-13	80	130			3
C-14	20	10			10
C-15	90	110			8
C-16	50	80			9
C-17	60	100			6
C-18	100	150			1

Table A.20 Responses of undergraduate (UG) students on the presented watershed management criteria

UG5					
Criteria	Q1	Q2	Q3	Q4	Q5
C-1	78	95	78	90	8
C-2	95	140	85	120	4
C-3	70	70	65	50	15
C-4	68	40	70	70	16
C-5	90	100	80	100	5
C-6	60	10	60	60	17
C-7			72	80	12
C-8			88	160	3
C-9			50	10	18
C-10			75	95	11
C-11			95	140	6
C-12			100	180	1
C-13	85	120			7
C-14	88	90			9
C-15	100	150			2
C-16	75	65			13
C-17	80	80			14
C-18	72	85			10

Table A.21 Responses of undergraduate (UG) students on the presented watershed management criteria

UG6					
Criteria	Q1	Q2	Q3	Q4	Q5
C-1	65	70	100	120	1
C-2	95	80	75	80	9
C-3	10	20	20	20	17
C-4	75	100	85	110	4
C-5	45	65	50	30	15
C-6	40	30	30	50	13
C-7			40	40	14
C-8			65	70	10
C-9			10	10	18
C-10			55	60	12
C-11			95	100	7
C-12			90	90	5
C-13	55	95			8
C-14	20	10			16
C-15	100	110			2
C-16	30	40			11
C-17	85	120			6
C-18	50	60			3

Table A.22 Responses of undergraduate (UG) students on the presented watershed management criteria

UG7					
Criteria	Q1	Q2	Q3	Q4	Q5
C-1	100	150	65	80	3
C-2	73	60	50	50	11
C-3	20	10	10	15	17
C-4	80	80	55	55	6
C-5	95	100	75	90	4
C-6	55	20	30	30	16
C-7			70	60	5
C-8			100	150	1
C-9			0	10	18
C-10			80	100	2
C-11			45	40	9
C-12			60	70	10
C-13	85	70			8
C-14	60	35			15
C-15	75	50			7
C-16	65	40			14
C-17	70	45			12
C-18	50	30			13

Table A.23 Responses of undergraduate (UG) students on the presented watershed management criteria

UG8					
Criteria	Q1	Q2	Q3	Q4	Q5
C-1	100	150	100	150	1
C-2	30	10	20	30	17
C-3	60	60	60	80	7
C-4	50	50	50	90	8
C-5	40	30	40	20	16
C-6	69	70	70	100	11
C-7			80	50	14
C-8			30	70	9
C-9			0	10	18
C-10			15	40	10
C-11			66	60	13
C-12			90	140	2
C-13	89	80			6
C-14	20	20			12
C-15	90	90			4
C-16	66	40			15
C-17	70	100			5
C-18	99	140			3

Table A.24 Responses of undergraduate (UG) students on the presented watershed management criteria

UG9					
Criteria	Q1	Q2	Q3	Q4	Q5
C-1	100	130	90	200	1
C-2	80	60	80	120	3
C-3	65	120	10	30	16
C-4	85	90	60	20	12
C-5	60	150	85	100	5
C-6	75	80	55	40	11
C-7			30	50	10
C-8			70	140	6
C-9			5	10	18
C-10			65	60	9
C-11			75	150	4
C-12			100	180	2
C-13	70	30			15
C-14	55	70			13
C-15	50	110			7
C-16	90	20			8
C-17	10	40			17
C-18	20	10			14

Table A.25 Responses of undergraduate (UG) students on the presented watershed management criteria

UG10					
Criteria	Q1	Q2	Q3	Q4	Q5
C-1	99	300	89	60	10
C-2	78	50	85	40	12
C-3	85	100	90	70	11
C-4	75	30	79	30	15
C-5	77	60	80	50	13
C-6	74	20	78	20	16
C-7			100	500	2
C-8			99	400	3
C-9			50	10	18
C-10			98	300	4
C-11			97	200	5
C-12			96	100	7
C-13	79	70			9
C-14	90	200			6
C-15	100	500			1
C-16	76	40			14
C-17	50	10			17
C-18	80	80			8

Table A.26 Responses of undergraduate (UG) students on the presented watershed management criteria

UG11					
Criteria	Q1	Q2	Q3	Q4	Q5
C-1	85	160	90	170	6
C-2	90	170	87	150	7
C-3	30	10	20	10	18
C-4	70	100	60	80	11
C-5	50	30	40	50	16
C-6	65	70	52	100	12
C-7			70	30	15
C-8			100	200	1
C-9			30	20	17
C-10			80	130	2
C-11			95	190	5
C-12			65	90	8
C-13	75	90			10
C-14	60	40			14
C-15	95	200			3
C-16	40	50			13
C-17	100	250			4
C-18	80	140			9

Table A.27 Responses of undergraduate (UG) students on the presented watershed management criteria

UG12					
Criteria	Q1	Q2	Q3	Q4	Q5
C-1	100	300	100	300	1
C-2	50	140	50	130	13
C-3	40	110	40	110	14
C-4	10	50	20	50	16
C-5	30	80	30	80	15
C-6	20	30	10	30	17
C-7			70	200	8
C-8			60	170	9
C-9			0	10	18
C-10			80	230	10
C-11			90	250	12
C-12			95	270	3
C-13	90	170			6
C-14	0	10			17
C-15	60	200			7
C-16	80	250			5
C-17	70	230			4
C-18	95	280			2

Table A.28 Responses of undergraduate (UG) students on the presented watershed management criteria

UG13					
Criteria	Q1	Q2	Q3	Q4	Q5
C-1	100	150	97	120	2
C-2	80	75	80	65	13
C-3	55	30	60	60	15
C-4	75	70	85	85	11
C-5	50	20	70	30	16
C-6	70	80	75	70	10
C-7			88	90	9
C-8			95	80	12
C-9			20	10	18
C-10			90	100	6
C-11			99	140	4
C-12			100	150	1
C-13	65	85			8
C-14	49	10			17
C-15	97	120			5
C-16	99	140			3
C-17	90	90			14
C-18	92	100			7

Table A.29 Responses of undergraduate (UG) students on the presented watershed management criteria

UG14					
Criteria	Q1	Q2	Q3	Q4	Q5
C-1	100	100	80	50	3
C-2	59	25	75	30	11
C-3	65	28	70	20	12
C-4	58	15	60	11	15
C-5	70	18	68	13	14
C-6	60	12	55	12	16
C-7			88	80	6
C-8			90	70	8
C-9			50	10	18
C-10			78	120	9
C-11			99	150	5
C-12			100	200	1
C-13	85	80			2
C-14	61	10			17
C-15	90	85			10
C-16	62	30			13
C-17	63	60			4
C-18	64	50			7

Table A.30 Responses of undergraduate (UG) students on the presented watershed management criteria

UG15					
Criteria	Q1	Q2	Q3	Q4	Q5
C-1	100	250	90	100	1
C-2	70	150	85	60	6
C-3	50	10	70	40	17
C-4	60	60	60	20	15
C-5	80	40	65	80	10
C-6	45	30	55	110	11
C-7			50	50	14
C-8			100	150	2
C-9			40	10	18
C-10			45	70	7
C-11			75	30	13
C-12			80	90	3
C-13	85	140			4
C-14	50	70			8
C-15	75	80			9
C-16	40	50			16
C-17	55	100			12
C-18	80	200			5

Appendix B

Survey raw data for postgraduate and undergraduate students' preferences on all watershed management criteria for all five questions separately.

Postgraduate Students

(See Tables B.1, B.2, B.3, B.4 and B.5).

Undergraduate Students

(See Tables B.6, B.7, B.8, B.9 and B.10).

N.H. Zardari et al., *Weighting Methods and their Effects on Multi-Criteria Decision Making Model Outcomes in Water Resources Management*,
SpringerBriefs in Water Science and Technology, DOI 10.1007/978-3-319-12586-2

Table B.1 Responses of postgraduate (PG) students for all five questions separately

Q1 for each respondent															
Criteria	PG1	PG2	PG3	PG4	PG5	PG6	PG7	PG8	PG9	PG10	PG11	PG12	PG13	PG14	PG15
C-1	90	100	100	100	100	100	100	45	80	100	100	70	85	80	100
C-2	80	95	80	85	85	60	65	28	55	40	2	40	70	90	90
C-3	70	40	80	50	30	90	80	80	50	50	12	55	90	50	70
C-4	50	70	70	75	50	50	75	20	70	70	60	80	100	60	95
C-5	50	60	65	60	99	70	92	70	85	30	20	40	65	10	85
C-6	50	50	10	70	60	20	85	30	60	65	70	65	80	30	75
C-7	Criteria 7 to Criteria 12 were not part of Question-1 of the survey questionnaire														
C-8															
C-9															
C-10															
C-11															
C-12															
C-13	60	75	95	65	80	5	76	50	75	80	10	85	60	40	90
C-14	60	55	70	60	90	15	45	35	45	45	1	40	20	75	50
C-15	100	80	100	80	97	45	60	40	90	45	4	65	40	95	80
C-16	80	70	20	75	70	5	70	100	65	60	0.5	95	45	35	95
C-17	90	88	90	70	96	55	90	60	95	55	30	65	15	100	40
C-18	70	90	100	90	98	80	50	65	100	90	15	100	50	80	95

Table B.2 Responses of postgraduate (PG) students for all five questions separately

Q2 for each respondent

Criteria	PG1	PG2	PG3	PG4	PG5	PG6	PG7	PG8	PG9	PG10	PG11	PG12	PG13	PG14	PG15
C-1	120	500	1980	1280	250	100	100	30	110	100	3000	70	360	100	200
C-2	100	400	960	640	110	65	80	10	40	30	50	20	120	80	190
C-3	50	10	240	10	50	90	70	35	20	45	100	30	540	50	40
C-4	50	120	960	80	30	30	60	15	50	70	1000	60	720	30	190
C-5	30	80	60	40	210	70	70	70	80	10	60	10	60	10	170
C-6	10	30	10	80	10	15	75	10	60	60	30	40	180	15	170
C-7	Criteria 7 to Criteria 12 were not part of Question-2 of the survey questionnaire														
C-8															
C-9															
C-10															
C-11															
C-12															
C-13	90	100	960	20	90	25	80	60	90	50	270	40	40	55	180
C-14	90	60	120	160	130	20	85	20	10	15	90	20	30	40	20
C-15	150	200	960	320	150	50	88	25	70	20	15	50	30	90	50
C-16	120	150	20	320	70	10	60	90	30	40	10	60	20	20	190
C-17	120	250	480	640	170	40	70	40	100	50	20	50	10	110	10
C-18	100	300	480	160	190	60	84	45	120	90	180	80	20	100	180

Table B.3 Responses of postgraduate (PG) students for all five questions separately

Q3 for each respondent															
Criteria	PG1	PG2	PG3	PG4	PG5	PG6	PG7	PG8	PG9	PG10	PG11	PG12	PG13	PG14	PG15
C-1	80	100	95	100	100	100	100	20	100	100	100	80	85	90	100
C-2	70	95	98	80	80	60	20	25	25	65	15	40	75	90	95
C-3	50	40	85	65	60	75	80	40	10	70	5	80	90	40	70
C-4	50	60	80	90	55	50	55	30	60	80	70	85	100	80	95
C-5	60	65	70	85	90	70	60	80	90	30	10	40	70	25	60
C-6	50	50	10	80	50	30	50	35	80	70	40	90	65	60	90
C-7	80	88	30	90	95	90	95	50	70	60	60	70	45	85	80
C-8	90	90	100	75	85	80	75	60	15	55	50	70	30	20	70
C-9	70	30	70	60	45	0	35	55	20	50	35	50	50	10	5
C-10	90	85	90	95	75	85	96	45	50	40	80	70	40	70	20
C-11	100	80	20	70	65	10	98	65	40	75	85	60	35	100	70
C-12	90	89	90	93	70	40	89	100	30	90	20	100	60	95	75
C-13	Criteria 13 to Criteria 18 were not part of Question-3 of the survey questionnaire														
C-14															
C-15															
C-16															
C-17															
C-18															

Table B.4 Responses of postgraduate (PG) students for all five questions separately

Q4 for each respondent															
Criteria	PG1	PG2	PG3	PG4	PG5	PG6	PG7	PG8	PG9	PG10	PG11	PG12	PG13	PG14	PG15
C-1	120	500	1280	100	250	100	120	10	100	100	14400	40	400	140	200
C-2	80	400	320	40	130	60	10	15	30	60	30	20	350	130	190
C-3	60	30	320	20	60	75	80	30	10	50	15	40	450	60	90
C-4	50	70	80	70	50	65	20	20	70	70	3600	80	500	100	190
C-5	50	100	120	30	220	70	60	70	100	10	10	10	300	20	25
C-6	10	50	10	70	20	40	30	25	90	65	360	90	250	40	180
C-7	100	200	40	90	190	95	100	40	80	25	400	50	160	110	150
C-8	80	280	640	50	160	80	40	50	15	30	180	50	40	30	30
C-9	60	10	50	10	10	10	50	45	20	40	60	30	80	10	10
C-10	150	180	640	50	110	90	90	35	60	20	200	50	20	80	20
C-11	150	150	20	60	70	25	70	55	50	80	7200	70	10	150	70
C-12	150	250	160	80	80	30	110	90	40	90	90	100	180	145	75
C-13	Criteria 13 to Criteria 18 were not part of Question-4 of the survey questionnaire														
C-14															
C-15															
C-16															
C-17															
C-18															

Table B.5 Responses of postgraduate (PG) students for all five questions separately

Q5 for each respondent (Rank order changed into Inverse Rank)															
Criteria	PG1	PG2	PG3	PG4	PG5	PG6	PG7	PG8	PG9	PG10	PG11	PG12	PG13	PG14	PG15
C-1	15	18	18	18	18	18	18	5	18	18	18	9	16	12	18
C-2	7	17	14	13	10	11	1	3	5	3	11	3	14	14	13
C-3	4	2	10	2	5	15	16	13	2	11	5	6	17	8	8
C-4	2	3	5	15	3	10	2	6	14	15	16	11	18	10	17
C-5	3	6	8	3	16	14	14	16	11	1	3	1	13	2	11
C-6	1	5	1	14	2	5	3	4	13	14	13	16	15	4	14
C-7	8	15	4	17	17	17	13	11	15	7	15	7	11	11	12
C-8	6	16	17	12	14	12	4	7	3	6	1	5	5	3	10
C-9	5	1	7	1	1	1	5	8	4	5	4	4	10	1	1
C-10	13	12	12	11	9	16	12	2	10	4	12	14	2	6	3
C-11	17	13	3	10	6	9	11	9	7	8	17	15	6	18	7
C-12	12	14	13	16	8	7	10	17	6	16	2	18	12	17	5
C-13	11	8	9	4	7	4	8	15	16	12	14	10	9	7	9
C-14	10	4	6	5	11	3	6	12	1	2	8	2	1	9	2
C-15	18	10	16	9	15	8	9	1	8	9	7	8	3	15	6
C-16	9	7	2	7	4	2	7	18	9	13	6	13	7	5	15
C-17	16	9	11	8	13	6	15	14	12	10	10	12	4	16	4
C-18	14	11	15	6	12	13	17	10	17	17	9	17	8	13	16

Table B.6 Responses of undergraduate (UG) students for all five questions separately

Q1 for each respondent															
Criteria	UG1	UG2	UG3	UG4	UG5	UG6	UG7	UG8	UG9	UG10	UG11	UG12	UG13	UG14	UG15
C-1	100	99	95	75	78	65	100	100	100	99	85	100	100	100	100
C-2	70	97	50	70	95	95	73	30	80	78	90	50	80	59	70
C-3	79	10	30	40	70	10	20	60	65	85	30	40	55	65	50
C-4	78	30	20	45	68	75	80	50	85	75	70	10	75	58	60
C-5	72	20	0	30	90	45	95	40	60	77	50	30	50	70	80
C-6	77	35	10	55	60	40	55	69	75	74	65	20	70	60	45
C-7	Criteria 7 to Criteria 12 were not part of Question-1 of the survey questionnaire														
C-8															
C-9															
C-10															
C-11															
C-12															
C-13	84	84	70	80	85	55	85	89	70	79	75	90	65	85	85
C-14	82	83	60	20	88	20	60	20	55	90	60	0	49	61	50
C-15	65	100	90	90	100	100	75	90	50	100	95	60	97	90	75
C-16	59	95	40	50	75	30	65	66	90	76	40	80	99	62	40
C-17	63	90	80	60	80	85	70	70	10	50	100	70	90	63	55
C-18	98	85	100	100	72	50	50	99	20	80	80	95	92	64	80

Table B.7 Responses of undergraduate (UG) students for all five questions separately

Q2 for each respondent															
Criteria	UG1	UG2	UG3	UG4	UG5	UG6	UG7	UG8	UG9	UG10	UG11	UG12	UG13	UG14	UG15
C-1	200	170	110	120	95	70	150	150	130	300	160	300	150	100	250
C-2	80	160	60	110	140	80	60	10	60	50	170	140	75	25	150
C-3	120	10	40	40	70	20	10	60	120	100	10	110	30	28	10
C-4	111	50	20	60	40	100	80	50	90	30	100	50	70	15	60
C-5	90	40	30	20	100	65	100	30	150	60	30	80	20	18	40
C-6	110	60	10	50	10	30	20	70	80	20	70	30	80	12	30
C-7	Criteria 7 to Criteria 12 were not part of Question-2 of the survey questionnaire														
C-8															
C-9															
C-10															
C-11															
C-12															
C-13	150	100	80	130	120	95	70	80	30	70	90	170	85	80	140
C-14	145	120	70	10	90	10	35	20	70	200	40	10	10	10	70
C-15	20	200	100	110	150	110	50	90	110	500	200	200	120	85	80
C-16	10	180	50	80	65	40	40	40	20	40	50	250	140	30	50
C-17	15	190	90	100	80	120	45	100	40	10	250	230	90	60	100
C-18	180	150	120	150	85	60	30	140	10	80	140	280	100	50	200

Table B.8 Responses of undergraduate (UG) students for all five questions separately

Q3 for each respondent															
Criteria	UG1	UG2	UG3	UG4	UG5	UG6	UG7	UG8	UG9	UG10	UG11	UG12	UG13	UG14	UG15
C-1	100	91	95	90	78	100	65	100	90	89	90	100	97	80	90
C-2	53	90	40	80	85	75	50	20	80	85	87	50	80	75	85
C-3	68	10	20	25	65	20	10	60	10	90	20	40	60	70	70
C-4	64	30	10	45	70	85	55	50	60	79	60	20	85	60	60
C-5	70	93	30	30	80	50	75	40	85	80	40	30	70	68	65
C-6	63	40	0	35	60	30	30	70	55	78	52	10	75	55	55
C-7	99	50	100	40	72	40	70	80	30	100	70	70	88	88	50
C-8	98	97	70	50	88	65	100	30	70	99	100	60	95	90	100
C-9	73	80	60	20	50	10	0	0	5	50	30	0	20	50	40
C-10	84	98	90	60	75	55	80	15	65	98	80	80	90	78	45
C-11	92	99	80	70	95	95	45	66	75	97	95	90	99	99	75
C-12	80	100	50	100	100	90	60	90	100	96	65	95	100	100	80
C-13	Criteria 13 to Criteria 18 were not part of Question-3 of the survey questionnaire														
C-14															
C-15															
C-16															
C-17															
C-18															

Table B.9 Responses of undergraduate (UG) students for all five questions separately

Q4 for each respondent															
Criteria	UG1	UG2	UG3	UG4	UG5	UG6	UG7	UG8	UG9	UG10	UG11	UG12	UG13	UG14	UG15
C-1	200	150	140	120	90	120	80	150	200	60	170	300	120	50	100
C-2	10	160	60	110	120	80	50	30	120	40	150	130	65	30	60
C-3	50	10	40	20	50	20	15	80	30	70	10	110	60	20	40
C-4	16	20	20	40	70	110	55	90	20	30	80	50	85	11	20
C-5	32	100	30	60	100	30	90	20	100	50	50	80	30	13	80
C-6	15	30	10	50	60	50	30	100	40	20	100	30	70	12	110
C-7	180	50	150	30	80	40	60	50	50	500	30	200	90	80	50
C-8	160	170	70	70	160	70	150	70	140	400	200	170	80	70	150
C-9	78	20	65	10	10	10	10	10	10	10	20	10	10	10	10
C-10	90	180	100	90	95	60	100	40	60	300	130	230	100	120	70
C-11	120	190	130	100	140	100	40	60	150	200	190	250	140	150	30
C-12	81	200	95	150	180	90	70	140	180	100	90	270	150	200	90
C-13	Criteria 13 to Criteria 18 were not part of Question-4 of the survey questionnaire														
C-14															
C-15															
C-16															
C-17															
C-18															

Table B.10 Responses of undergraduate (UG) students for all five questions separately

Q5 for each respondent (Rank order changed into Inverse Rank)

Criteria	UG1	UG2	UG3	UG4	UG5	UG6	UG7	UG8	UG9	UG10	UG11	UG12	UG13	UG14	UG15
C-1	18	9	16	15	11	18	16	18	18	9	13	18	17	16	18
C-2	4	12	8	14	15	10	8	2	16	7	12	6	6	8	13
C-3	8	1	3	2	4	2	2	12	3	8	1	5	4	7	2
C-4	6	3	2	4	3	15	13	11	7	4	8	3	8	4	4
C-5	7	6	4	3	14	4	15	3	14	6	3	4	3	5	9
C-6	5	2	1	6	2	6	3	8	8	3	7	2	9	3	8
C-7	16	5	18	5	7	5	14	5	9	17	4	11	10	13	5
C-8	15	14	7	8	16	9	18	10	13	16	18	10	7	11	17
C-9	9	4	5	1	1	1	1	1	1	1	2	1	1	1	1
C-10	11	11	11	12	8	7	17	9	10	15	17	9	13	10	12
C-11	14	17	12	7	13	12	10	6	15	14	14	7	15	14	6
C-12	10	18	10	17	18	14	9	17	17	12	11	16	18	18	16
C-13	13	7	13	16	12	11	11	13	4	10	9	13	11	17	15
C-14	12	8	9	9	10	3	4	7	6	13	5	2	2	2	11
C-15	3	16	15	11	17	17	12	15	12	18	16	12	14	9	10
C-16	1	13	6	10	6	8	5	4	11	5	6	14	16	6	3
C-17	2	15	14	13	5	13	7	14	2	2	15	15	5	15	7
C-18	17	10	17	18	9	16	6	16	5	11	10	17	12	12	14

Appendix C

Number of retrievals for each weighting method in combination of MCDM, MCDA, MCA, and MADM keywords for scopus and sciverse databases.

Scopus Database

(See Tables C.1, C.2 and Figs. C.1, C.2, C.3 C.4, C.5, C.6, C.7, C.8, C.9, C.10, C.11, C.12, C.13, C.14, C.15, C.16, C.17 and C.18).

N.H. Zardari et al., *Weighting Methods and their Effects on Multi-Criteria Decision Making Model Outcomes in Water Resources Management*,
SpringerBriefs in Water Science and Technology, DOI 10.1007/978-3-319-12586-2

Table C.1 Weighting methods and number of retrievals for all keywords (MCDM, MCDA, MCA, MADM) (scopus database only)

Rank order (based on last column values)	Weighting Method	Number of retrievals for specific weighting method and keyword [searching period = 13 years (2000–2012)]				Column Sum
		& MCDM	& MCDA	& MCA	& MADM	
1	Ranking	1383	440	122	354	**2299**
2	AHP	1205	275	45	277	**1802**
3	TOPSIS	876	100	11	343	**1330**
4	Pairwise Comparison	321	116	18	55	**510**
5	SMART	185	88	171	40	**484**
6	Trade-off	161	53	63	16	**293**
7	Delphi Method	90	9	7	24	**130**
8	SWING	8	15	44	2	**69**
9	Least Square Method	31	9	23	2	**65**
10	Entropy Method	39	3	2	21	**65**
11	Simos Procedure	28	26	2	5	**61**
12	Eigenvector Method	14	4	7	5	**30**
	Row sum	**4341**	**1138**	**515**	**1144**	**7138**

Table C.2 Journal versus number of search retrievals (scopus database only)

Rank order	Journal	Total retrievals (period 2000–2012)	Weight (%)
1	Expert Systems with Applications	488	21.7
2	European Journal of Operational Research	314	14.0
3	International Journal of Information Technology and Decision Making	95	4.2
4	Information Sciences	90	4.0
5	Omega	85	3.8
6	International Journal of Production Research	80	3.6
7	Materials and Design	72	3.2
8	Technological and Economic Development of Economy	68	3.0
9	International Journal of Advanced Manufacturing Technology	65	2.9
10	Journal of the Operational Research Society	64	2.9
11	Journal of Environmental Management	60	2.7
12	Applied Soft Computing Journal	57	2.5
13	Applied Mathematical Modelling	50	2.2
14	Computers and Industrial Engineering	47	2.1
15	Mathematical and Computer Modelling	47	2.1
16	Annals of Operations Research	43	1.9
17	Decision Support Systems	41	1.8
18	Computers and Operations Research	39	1.7
19	International Journal of Computational Intelligence Systems	39	1.7
20	International Journal of Production Economics	39	1.7
21	European Journal of Scientific Research	38	1.7
22	Knowledge Based Systems	37	1.6
23	Pesquisa Operacional	36	1.6
24	Energy Policy	34	1.5
25	Ecological Economics	33	1.5
26	Australian Journal of Basic and Applied Sciences	30	1.3
27	Kongzhi Yu Juece Control and Decision	28	1.2
28	International Journal of Services and Operations Management	27	1.2
29	Water Resources Management	27	1.2
30	Automation in Construction	24	1.1
31	Fuzzy Sets and Systems	24	1.1
32	Studies in Fuzziness and Soft Computing	23	1.0

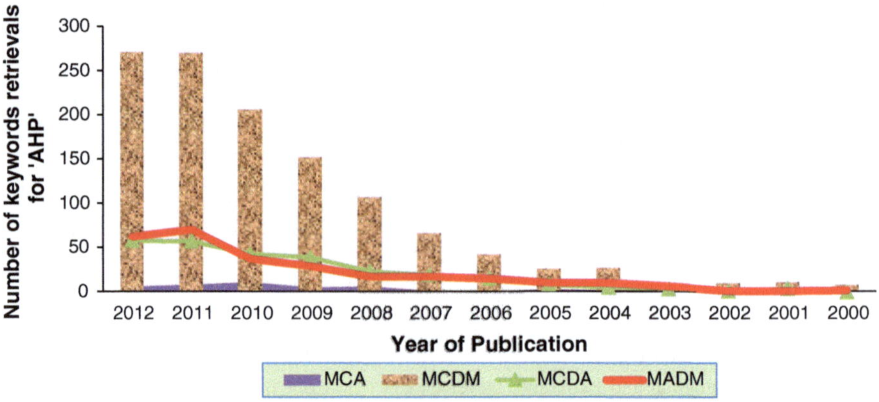

Fig. C.1 Number of keywords (MCDM, MCDA, MCA, MADM) retrievals in combination of "AHP" (scopus database only)

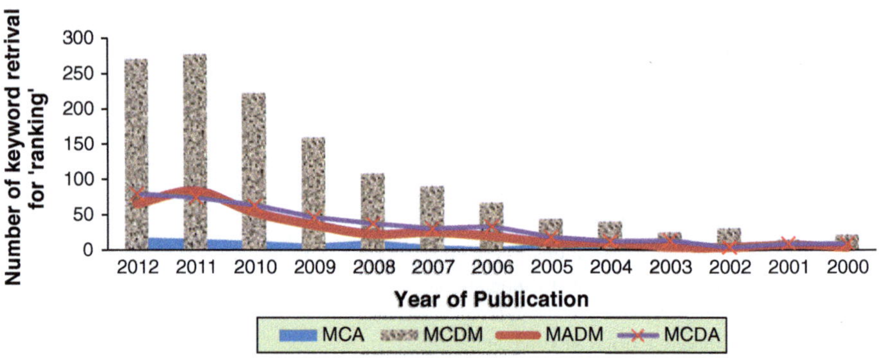

Fig. C.2 Number of keywords (MCDM, MCDA, MCA, MADM) retrievals in combination of "Ranking" (scopus database only)

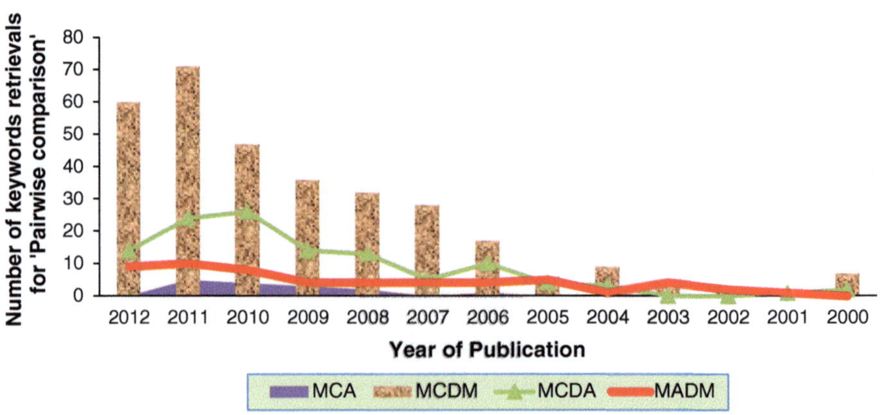

Fig. C.3 Number of keywords (MCDM, MCDA, MCA, MADM) retrievals in combination of "Pairwise comparison" (scopus database only)

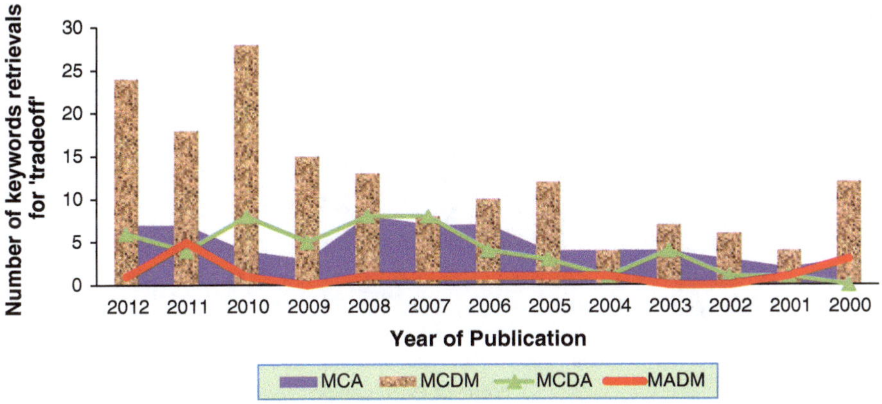

Fig. C.4 Number of keywords (MCDM, MCDA, MCA, MADM) retrievals in combination of "Trade-off" (scopus database only)

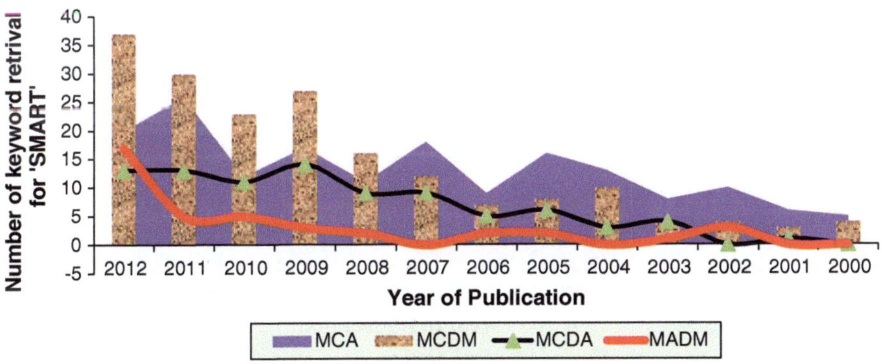

Fig. C.5 Number of keywords (MCDM, MCDA, MCA, MADM) retrievals in combination of "SMART" (scopus database only)

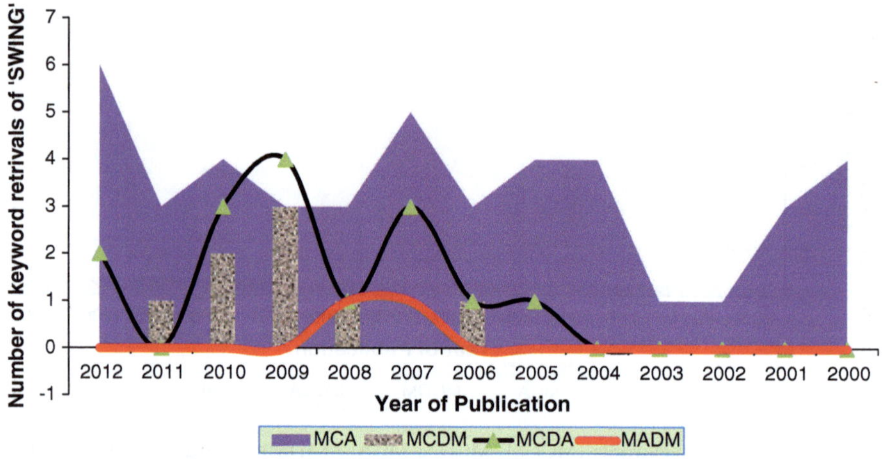

Fig. C.6 Number of keywords (MCDM, MCDA, MCA, MADM) retrievals in combination of "SWING" (scopus database only)

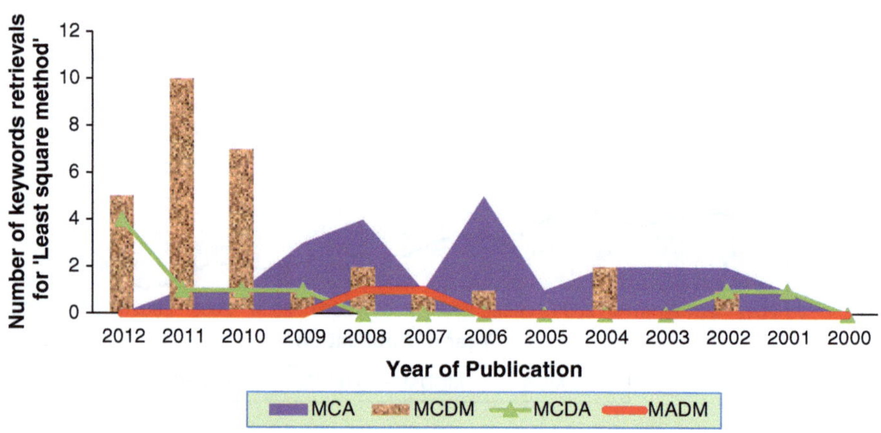

Fig. C.7 Number of keywords (MCDM, MCDA, MCA, MADM) retrievals in combination of "Least square method" (scopus database only)

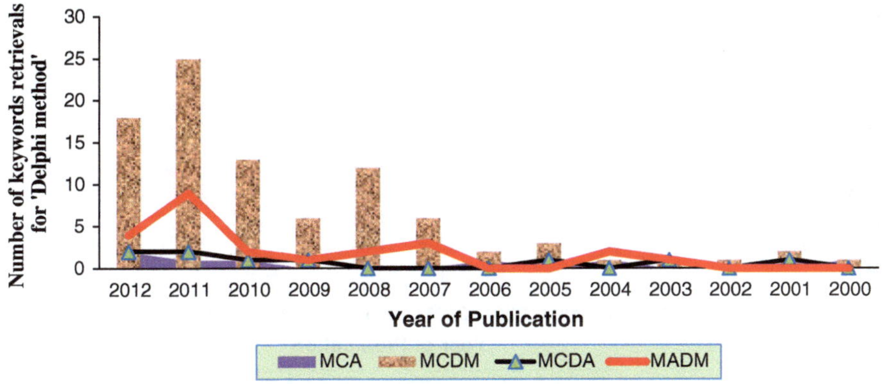

Fig. C.8 Number of keywords (MCDM, MCDA, MCA, MADM) retrievals in combination of "Delphi method" (scopus database only)

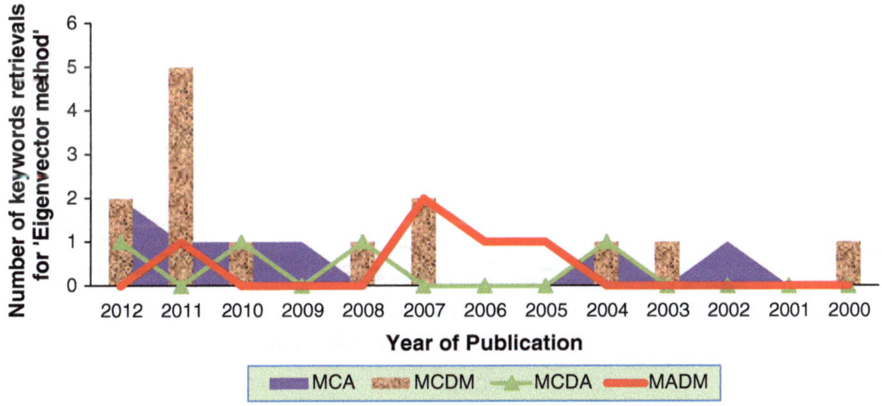

Fig. C.9 Number of keywords (MCDM, MCDA, MCA, MADM) retrievals in combination of "Eigenvector method" (scopus database only)

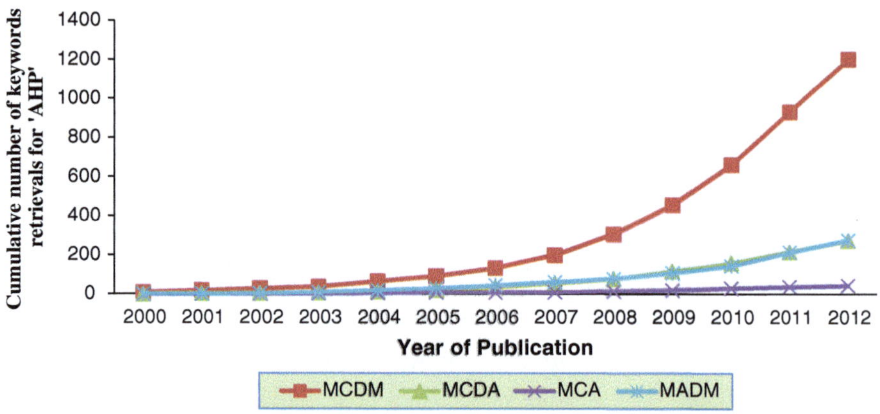

Fig. C.10 Cumulative plot of keywords (MCDM, MCDA, MCA, MADM) retrievals in combination of "AHP" keyword (scopus database only)

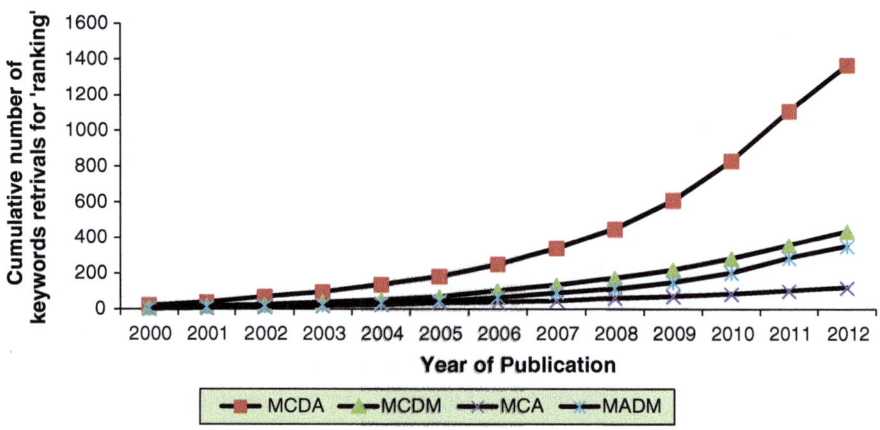

Fig. C.11 Cumulative plot of keywords (MCDM, MCDA, MCA, MADM) retrievals in combination of "Ranking" keyword (scopus database only)

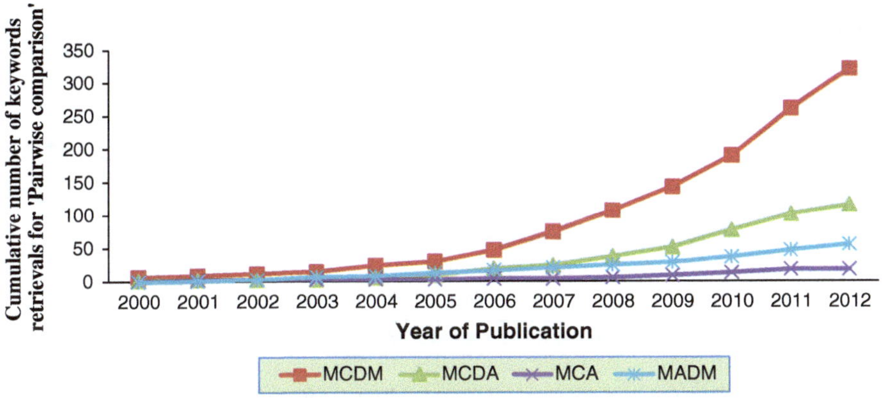

Fig. C.12 Cumulative plot of keywords (MCDM, MCDA, MCA, MADM) retrievals in combination of "Pairwise comparison" keyword (scopus database only)

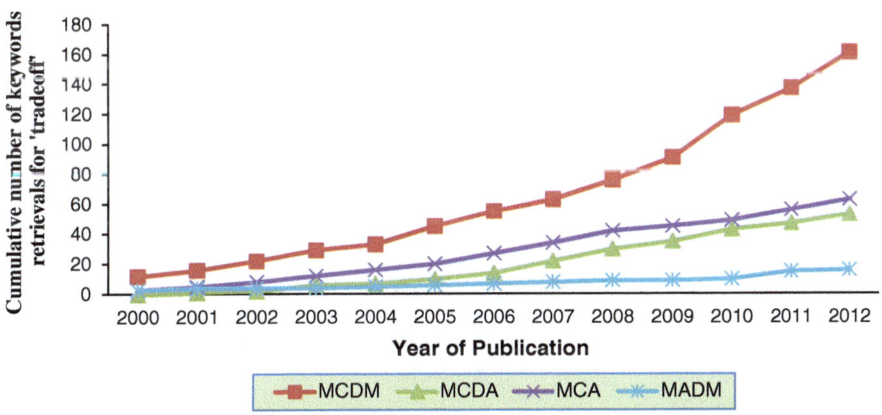

Fig. C.13 Cumulative plot of keywords (MCDM, MCDA, MCA, MADM) retrievals in combination of "Trade-off" keyword (scopus database only)

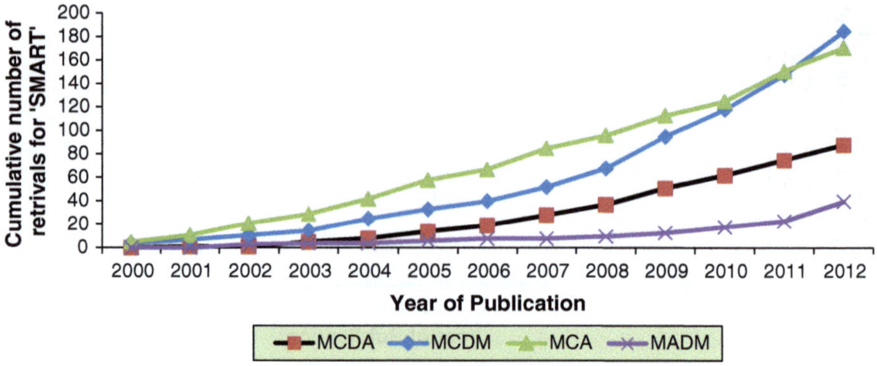

Fig. C.14 Cumulative plot of keywords (MCDM, MCDA, MCA, MADM) retrievals in combination of "SMART" keyword (scopus database only)

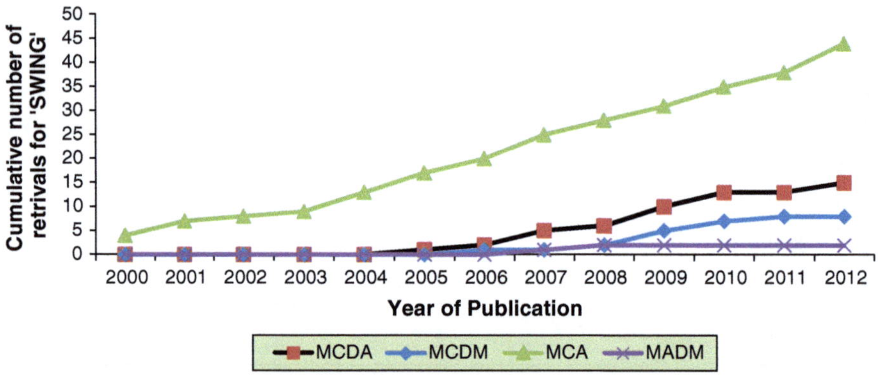

Fig. C.15 Cumulative plot of keywords (MCDM, MCDA, MCA, MADM) retrievals in combination of "SWING" keyword (scopus database only)

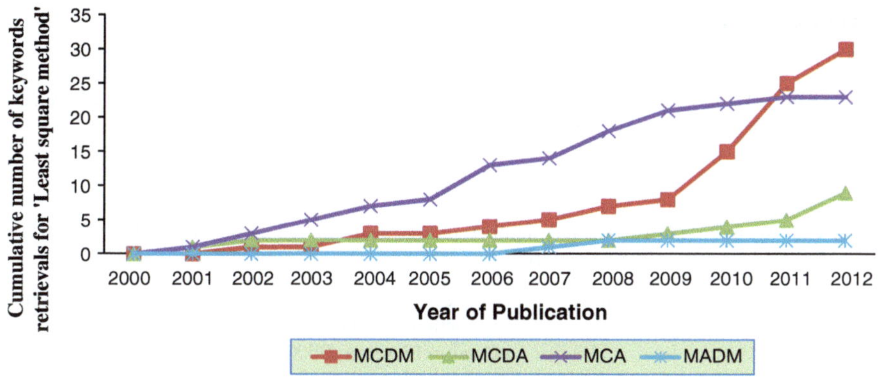

Fig. C.16 Cumulative plot of keywords (MCDM, MCDA, MCA, MADM) retrievals in combination of "Least square method" keyword (scopus database only)

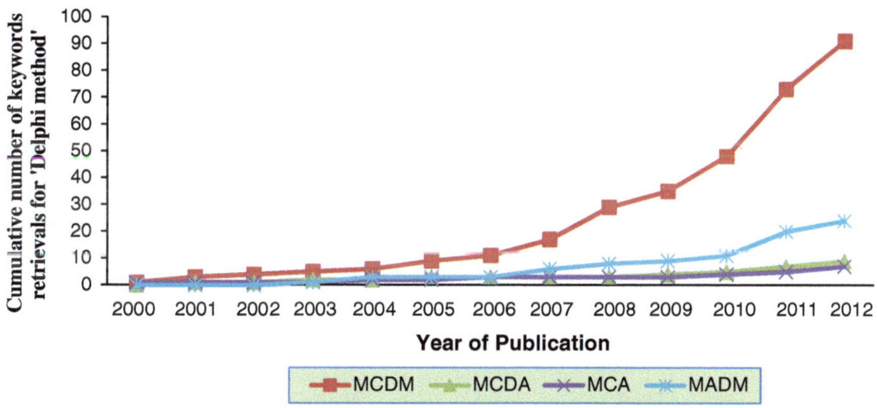

Fig. C.17 Cumulative plot of keywords (MCDM, MCDA, MCA, MADM) retrievals in combination of "Delphi" keyword (scopus database only)

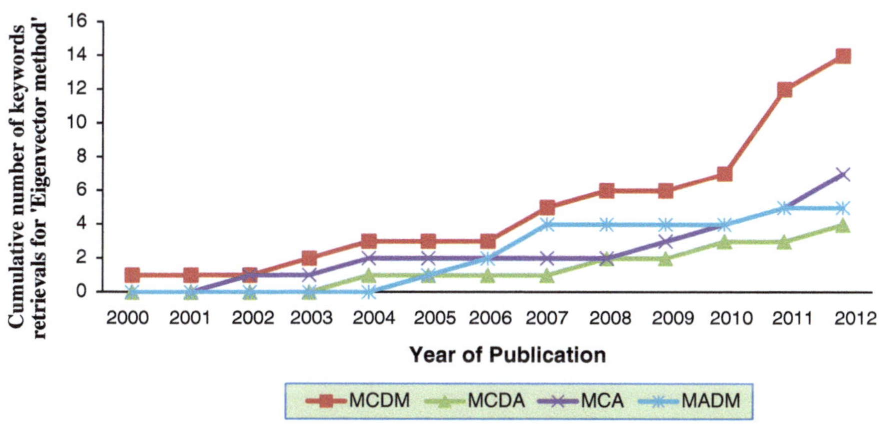

Fig. C.18 Cumulative plot of keywords (MCDM, MCDA, MCA, MADM) retrievals in combination of "Eigenvector method" keyword (scopus database only)

Sciverse Database

(See Tables C.3, C.4 and Figs. C.19, C.20, C.21, C.22, C.23, C.24, C.25, C.26, C.27, C.28, C.29, C.30, C.31, C.32, C.33, C.34, C.35, and C.36).

Table C.3 Weighting methods and number of retrievals for all keywords (MCDM, MCDA, MCA, MADM) [sciverse database only]

Rank order (based on last column values)	Weighting Method	Number of retrievals for specific weighting method and keyword [searching period = 13 years (2000–2012)]				Column Sum
		& MCDM	& MCDA	& MCA	& MADM	
1	Ranking	1758	766	634	548	**3706**
2	AHP	1331	407	164	388	**2290**
3	TOPSIS	738	128	18	372	**1256**
4	Trade-off	492	250	381	97	**1220**
5	Pairwise Comparison	569	201	133	135	**1038**
6	SMART	266	131	474	79	**950**
7	SWING	58	68	169	13	**308**
8	Delphi Method	133	32	26	44	**235**
9	Eigenvector Method	100	24	9	38	**171**
10	Entropy Method	76	9	14	48	**147**
11	Least Square Method	32	3	29	13	**77**
12	Simos Procedure	24	20	4	12	**60**
	Row sum	**5577**	**2039**	**2055**	**1787**	**11458**

Table C.4 Journal versus number of search retrievals (sciverse database only)

Rank order	Journal	Total retrievals (period 2000–2012)	Weight (%)
1	Expert Systems with Applications	1396	20.49
2	European Journal of Operational Research	1191	17.48
3	Mathematical and Computer Modelling	319	4.68
4	Omega	266	3.90
5	Journal of Environmental Radioactivity	261	3.83
6	Applied Soft Computing	250	3.67
7	Information Sciences	242	3.55
8	Computers and Operations Research	235	3.45
9	Decision Support Systems	234	3.44
10	Energy Policy	223	3.27
11	International Journal of Production Economics	187	2.75
12	Computers and Industrial Engineering	179	2.63
13	Ecological Economics	171	2.51
14	Knowledge-Based Systems	166	2.44
15	Applied Mathematical Modelling	163	2.39
16	Materials and Design	160	2.35
17	Environmental Modelling and Software	131	1.92
18	Journal of Cleaner Production	88	1.29
19	Fuzzy Sets and Systems	87	1.28
20	Renewable and Sustainable Energy Reviews	82	1.20
21	Science of The Total Environment	79	1.16
22	Automation in Construction	69	1.01
23	Applied Mathematics and Computation	61	0.90
24	Fuel and Energy Abstracts	58	0.85
25	Energy	57	0.84
26	Land Use Policy	55	0.81
27	Nuclear Instruments and Methods in Physics Research	55	0.81
28	Waste Management	48	0.70
29	Forest Ecology and Management	44	0.65
30	Ecological Indicators	37	0.54
31	Environmental Impact Assessment Review	35	0.51
32	Brain Research	33	0.48
33	Computer Communications	33	0.48
34	Procedia—Social and Behavioral Sciences	30	0.44
35	International Journal of Project Management	29	0.43
36	Value in Health	27	0.40
37	Building and Environment	15	0.22
38	Biological Conservation	12	0.18
39	Ecological Modelling	4	0.06

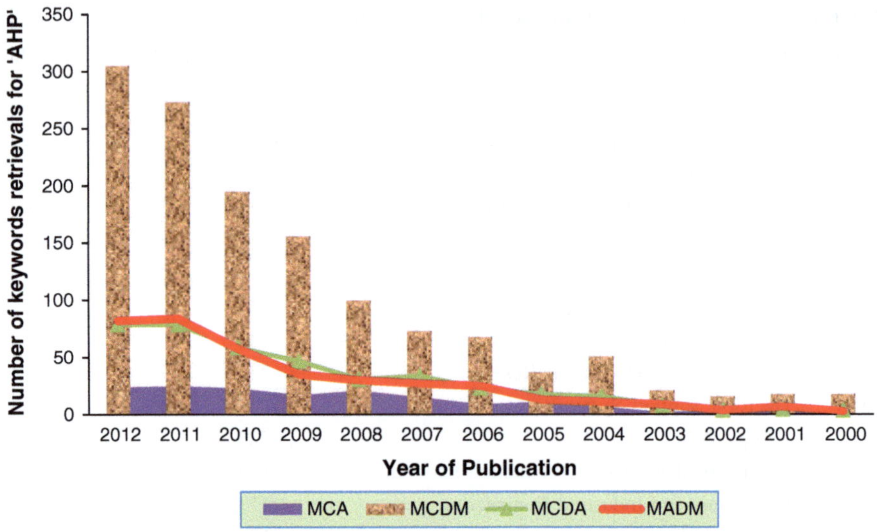

Fig. C.19 Number of keywords (MCDM, MCDA, MCA, MADM) retrievals in combination of "AHP" (sciverse database only)

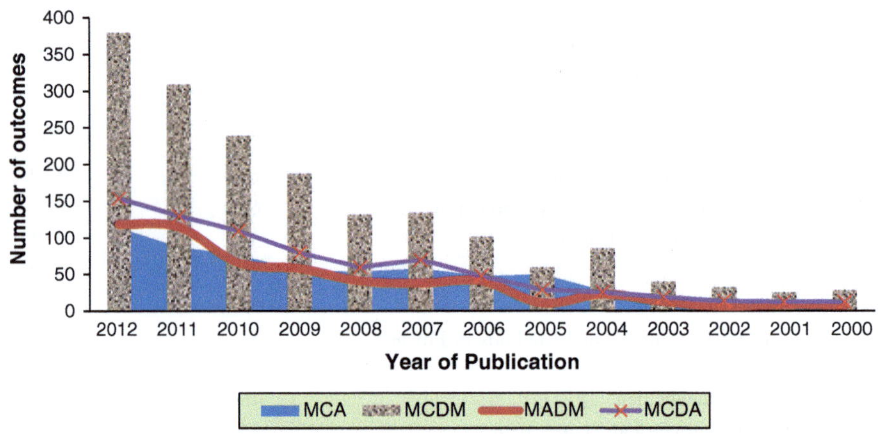

Fig. C.20 Number of keywords (MCDM, MCDA, MCA, MADM) retrievals in combination of "Ranking" (sciverse database only)

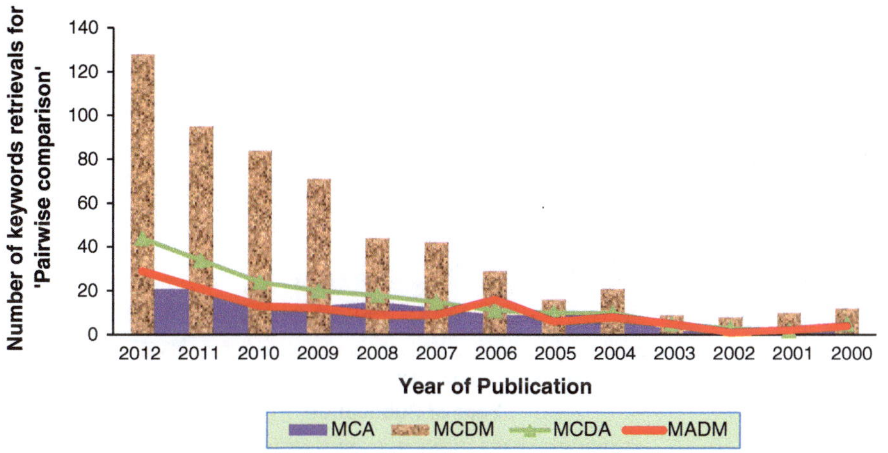

Fig. C.21 Number of keywords (MCDM, MCDA, MCA, MADM) retrievals in combination of "Pairwise comparison" (sciverse database only)

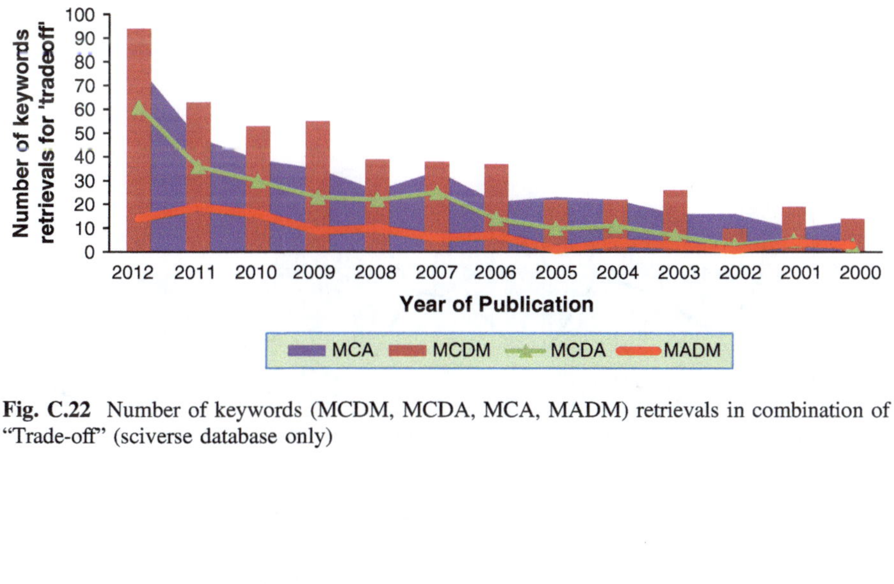

Fig. C.22 Number of keywords (MCDM, MCDA, MCA, MADM) retrievals in combination of "Trade-off" (sciverse database only)

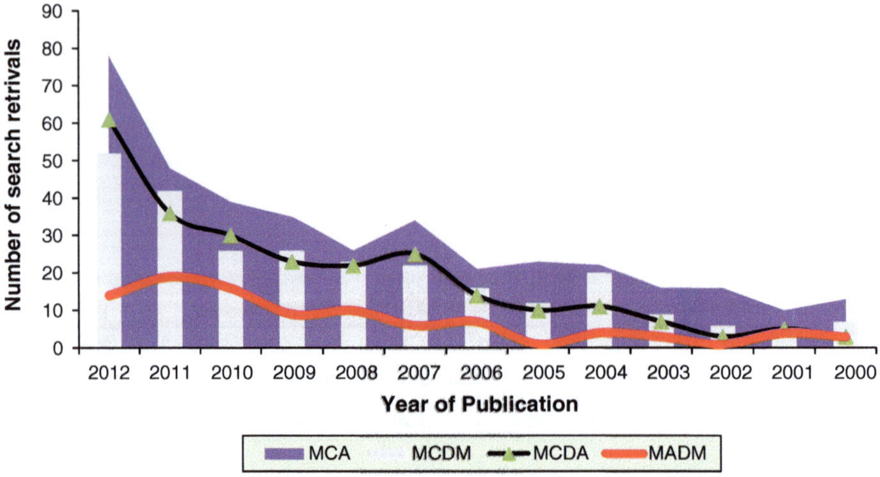

Fig. C.23 Number of keywords (MCDM, MCDA, MCA, MADM) retrievals in combination of "SMART" (sciverse database only)

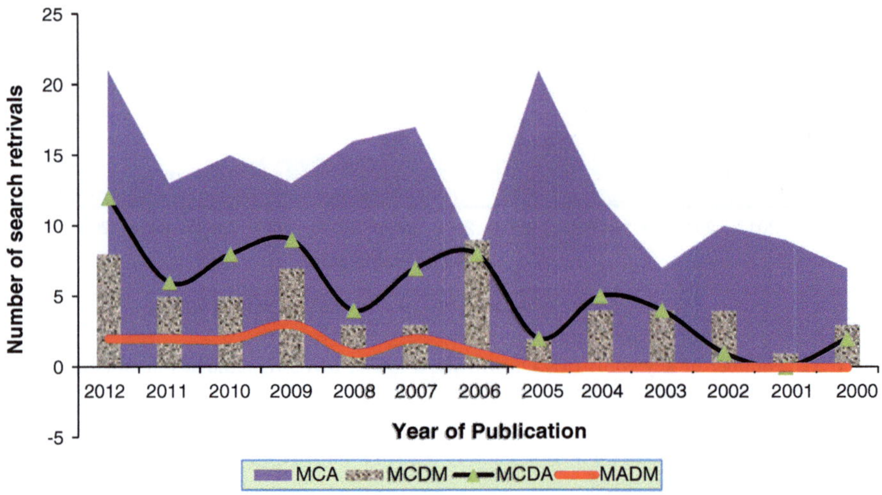

Fig. C.24 Number of keywords (MCDM, MCDA, MCA, MADM) retrievals in combination of "SWING" (sciverse database only)

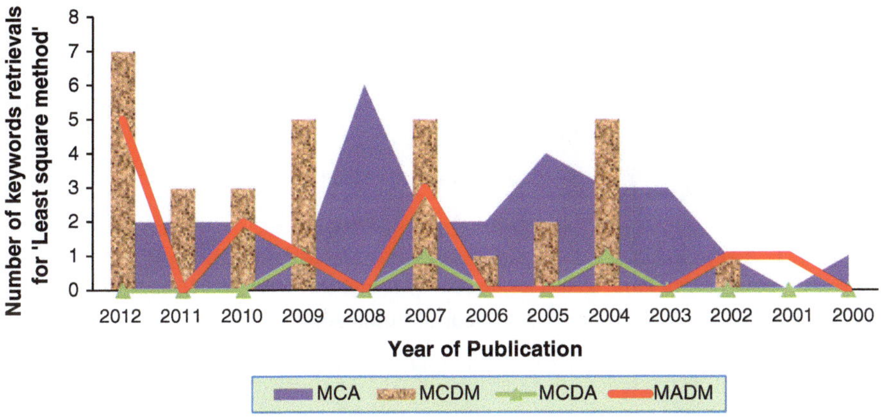

Fig. C.25 Number of keywords (MCDM, MCDA, MCA, MADM) retrievals in combination of "Least square method" (sciverse database only)

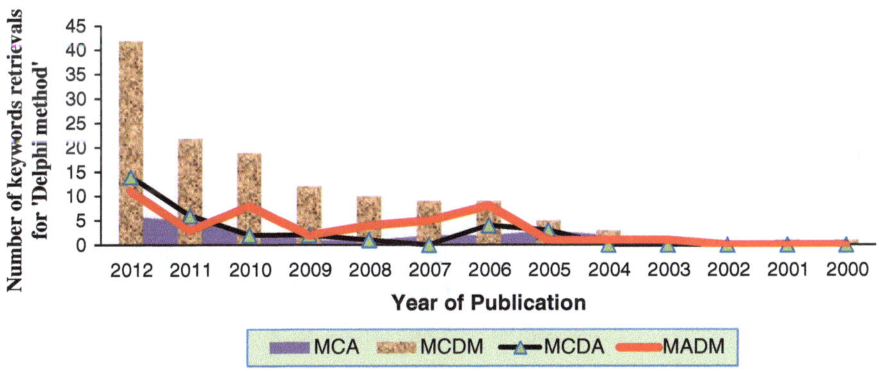

Fig. C.26 Number of keywords (MCDM, MCDA, MCA, MADM) retrievals in combination of "Delphi method" (sciverse database only)

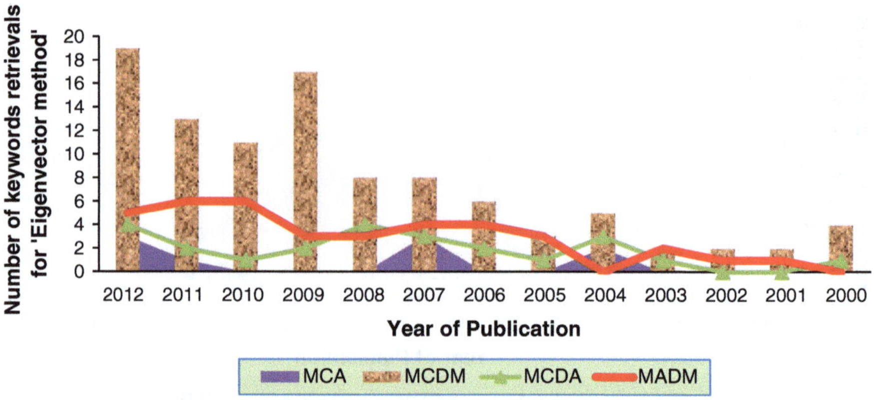

Fig. C.27 Number of keywords (MCDM, MCDA, MCA, MADM) retrievals in combination of "Eigenvector method" (sciverse database only)

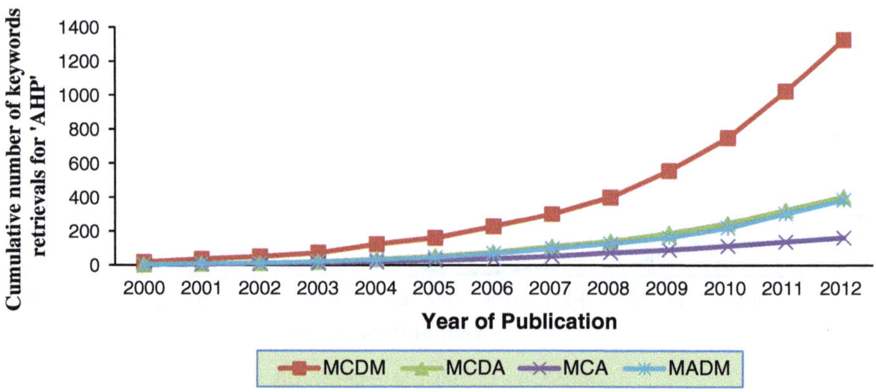

Fig. C.28 Cumulative plot of keywords (MCDM, MCDA, MCA, MADM) in combination with "AHP" keyword (sciverse database only)

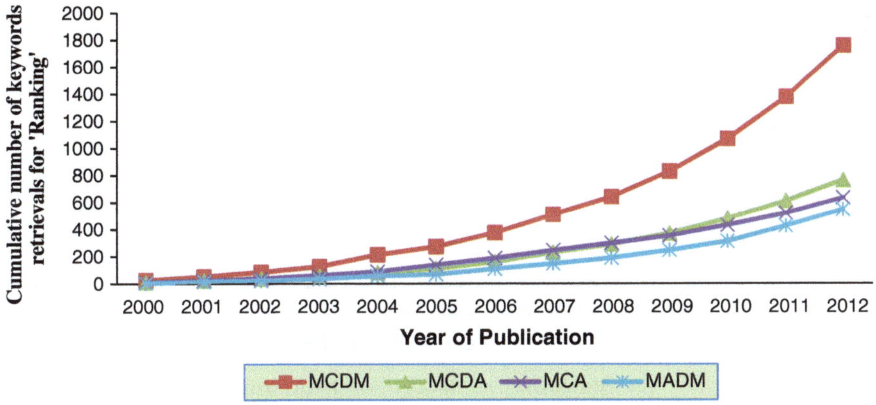

Fig. C.29 Cumulative plot of keywords (MCDM, MCDA, MCA, MADM) in combination with "Ranking" keyword (sciverse database only)

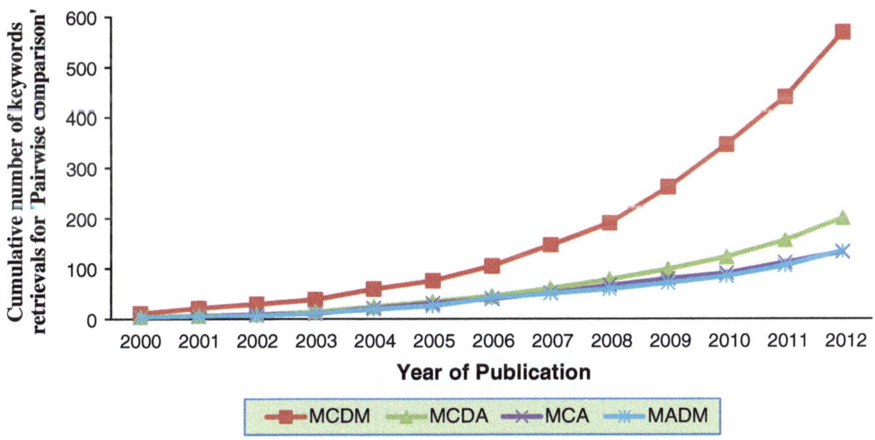

Fig. C.30 Cumulative plot of keywords (MCDM, MCDA, MCA, MADM) in combination with "Pairwise comparison" keyword (sciverse database only)

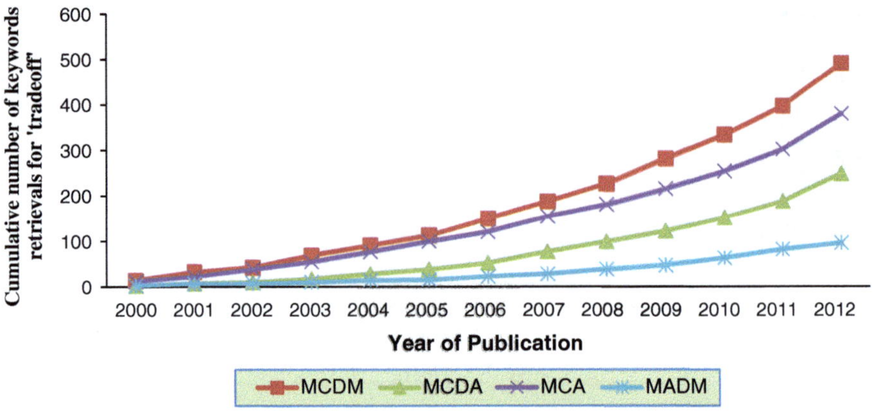

Fig. C.31 Cumulative plot of keywords (MCDM, MCDA, MCA, MADM) in combination with "Trade-off" keyword (sciverse database only)

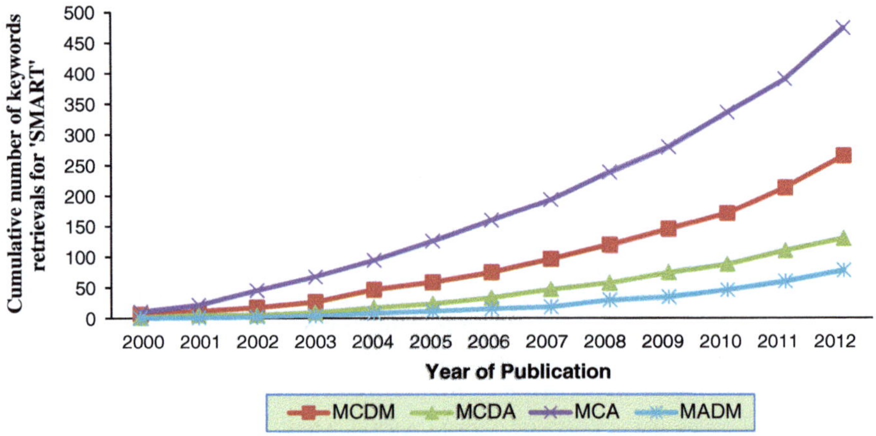

Fig. C.32 Cumulative plot of keywords (MCDM, MCDA, MCA, MADM) in combination with "SMART" keyword (sciverse database only)

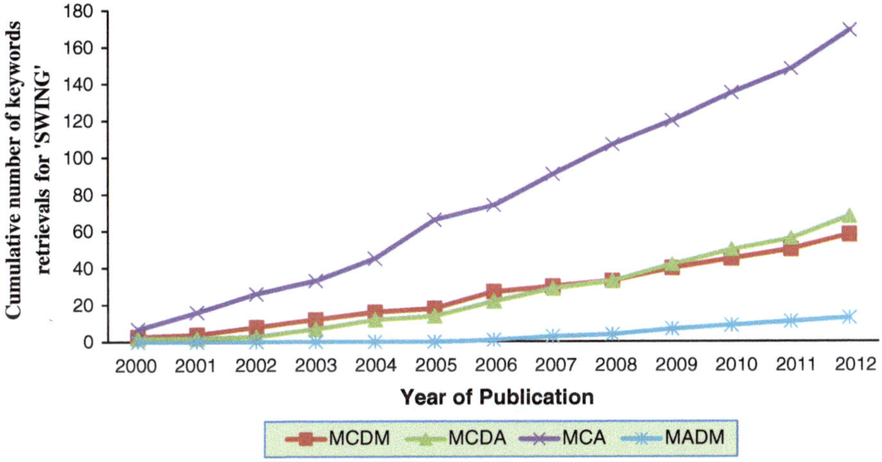

Fig. C.33 Cumulative plot of keywords (MCDM, MCDA, MCA, MADM) in combination with "SWING" keyword (sciverse database only)

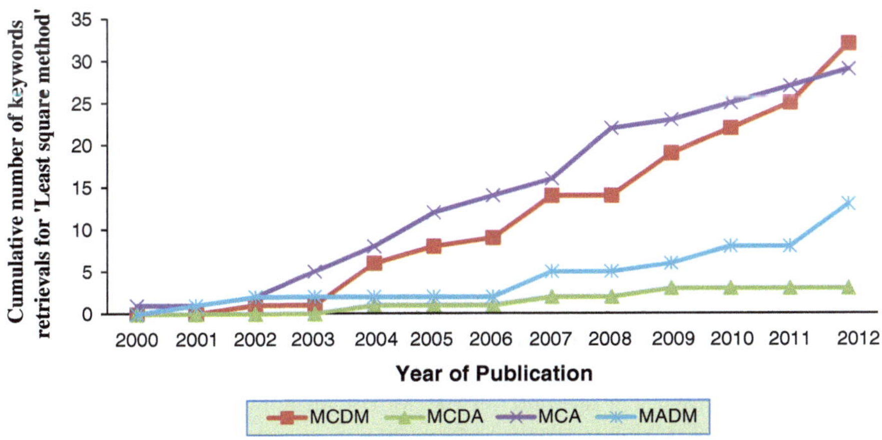

Fig. C.34 Cumulative plot of keywords (MCDM, MCDA, MCA, MADM) in combination with "Least square method" keyword (sciverse database only)

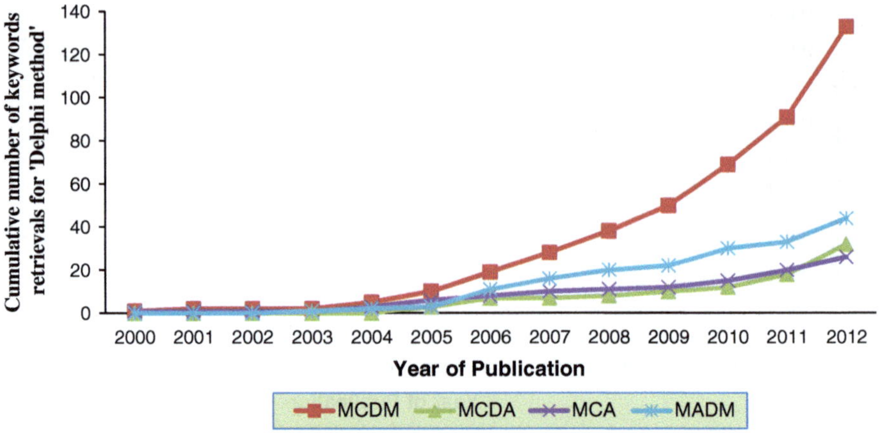

Fig. C.35 Cumulative plot of keywords (MCDM, MCDA, MCA, MADM) in combination with "Delphi" keyword (sciverse database only)

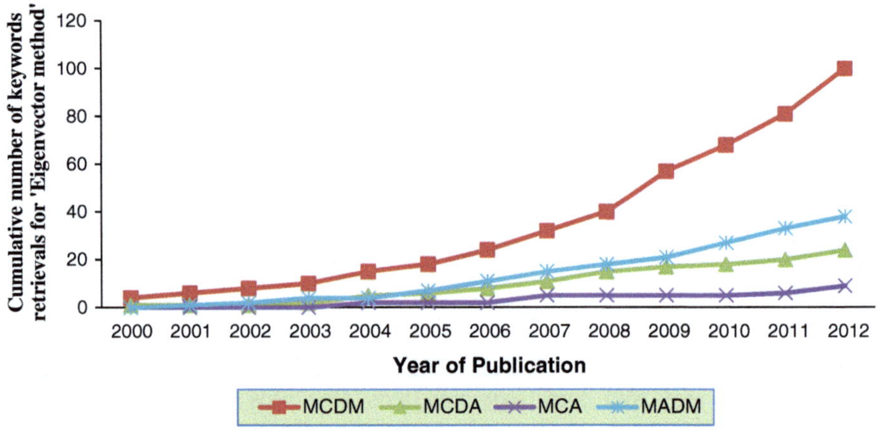

Fig. C.36 Cumulative plot of keywords (MCDM, MCDA, MCA, MADM) in combination with "Eigenvector method" keyword (sciverse database only)

References

Aalianvari, A., et al. (2012). Application of fuzzy Delphi AHP method for the estimation and classification of Ghomrud tunnel from groundwater flow hazard. *Arab Journal of Geoscience, 5*(2), 275–284.

Abrishamchi, A., Ebrahimian, A., Tajrishi, M., & Marino, M. A. (2005). Case study: application of multi-criteria decision making to urban water supply. *Journal of Water Resources Planning and Management, 131*(4), 326–335.

Afshar, A., Mariño, M. A., & Saadatpour, M. (2011). Fuzzy TOPSIS multi-criteria decision analysis applied to Karun reservoirs system. *Water Resources Management, 25*(2), 545–563.

Ahn, B. S. (2011). Compatible weighting method with rank order centroid: Maximum entropy ordered weighted averaging approach. *European Journal of Operational Research, 212*(3), 552–559.

Alipour, M. H., Shamsai, A., & Ahmady, N. (2010). A new fuzzy multi-criteria decision making method and its application in diversion of water. *Expert Systems with Applications, 37*(12), 8809–8813.

Al-Juaidi, A. E., Kaluarachchi, J. J., & Kim, U. (2010). Multi-Criteria decision analysis of treated wastewater use for agriculture in water deficit regions. *JAWRA Journal of the American Water Resources Association, 46*(2), 395–411.

Anagnostopoulos, K. P., & Petalas, C. (2011). A fuzzy multi-criteria benefit-cost approach for irrigation projects evaluation. *Agricultural Water Management, 98*(9), 1409–1416.

Arnette, A., et al. (2010). Stakeholder ranking of watershed goals with the vector analytic hierarchy process: Effects of participant grouping scenarios. *Environmental Modelling and Software, 25*(11), 1459–1469.

Babaei, H., Araghinejad, S., & Hoorfar, A. (2012). Developing a new method for spatial assessment of drought vulnerability (case study: Zayandeh-Rood river basin in Iran). *Water and Environment Journal*, (in press).

Balasubramaniam, A., Boyle, A. R., & Voulvoulis, N. (2007). Improving petroleum contaminated land remediation decision-making through the MCA weighting process. *Chemosphere, 66*(5), 791–798.

Bana e Costa, C. A., Antão da Silva, P., & Nunes Correia, F. (2004). Multicriteria evaluation of flood control measures: The case of Ribeirado Livramento. *Water Resources Management, 18* (3), 263–283.

Beinat, E. (1997). *Environmental valuation functions*. Dordrecht: Kluwer.

Bello-Dambatta, A., et al. (2009). The analytical hierarchy process for contaminated land management. *Advanced Engineering Informatics, 23*(4), 433–441.

Belton, V., & Stewart, T. J. (2002). *Multiple criteria decision analysis: An integrated approach.* Dordrecht: Kluwer Academic Publishing.

Benzerra, A., et al. (2012). Decision support for sustainable urban drainage system management: A case study of Jijel. *Algeria, Journal of Environmental Management, 101*, 46–53.

N.H. Zardari et al., *Weighting Methods and their Effects on Multi-Criteria Decision Making Model Outcomes in Water Resources Management*,
SpringerBriefs in Water Science and Technology, DOI 10.1007/978-3-319-12586-2

Biswas, S., et al. (2012). Evaluating integrated watershed management using multiple criteria analysis—a case study at Chittagong Hill Tracts in Bangladesh. *Environmental Monitoring and Assessment, 184*(5), 2741–2761.

Bobylev, N. (2009). In T. H. Illangasekare, K. Mahutova & J. J. Barich (Eds.), *Multiple criteria decision making and environmental security decision support for natural disasters and intentional threats to water security* (pp. 213–228). Netherlands: Springer.

Brans, J. P., & Vincke, Ph. (1985). A preference ranking organisation method: (The PROMETHEE method for multiple criteria decision-making). *Management Science, 31*(6), 647–656.

Bryan, B. A. (2010). Development and application of a model for robust, cost-effective investment in natural capital and ecosystem services. *Biological Conservation, 143*(7), 1737–1750.

Calizaya, A., et al. (2010). Multi-criteria decision analysis (MCDA) for integrated water resources management (IWRM) in the Lake Poopo Basin, Bolivia. *Water Resources Management, 24* (10), 2267–2289.

Chang, N. B., et al. (2010). Development of the metropolitan water availability index (MWAI) and short-term assessment with multi-scale remote sensing technologies. *Journal of Environmental Management, 91*(6), 1397–1413.

Chang, Y. H., & Yeh, C. H. (2001). Evaluating airline competitiveness using multiattribute decision making. *Omega, 29*(5), 405–415.

Chen, S. Y., & Hou, Z. C. (2004). Multi-criterion decision making for flood control operations: Theory and applications. *Journal of the American Water Resources Association, 40*(1), 67–76.

Chen, V. Y., et al. (2011). Fuzzy MCDM approach for selecting the best environment-watershed plan. *Applied Soft Computing,11*(1), 265–275.

Chen, Y., Okudan, G. E., & Riley, D. R. (2010). Decision support for construction method selection in concrete buildings: Prefabrication adoption and optimization. *Automation in Construction, 19*(6), 665–675.

Chen, Y. C., et al. (2009). Fuzzy MCDM technique for planning the environment watershed cutting-edge research topics on multiple criteria decision making. In Y. Shi et al. (Eds.), (pp. 744–752). Berlin, Heidelberg: Springer.

Chen, Z. J. & Zhang Y. (2010). *Application of TOPSIS analysis method based on AHP in bid evaluation of power equipment*. 2nd IEEE International Conference on Information and Financial Engineering (ICIFE), 17–19 September, Chongqing, China, pp. 193–196.

Choo, E. U., Schoner, B., & Wedley, W. C. (1999). Interpretation of criteria weights in multi-criteria decision making. *Computers and Industrial Engineering, 37*, 527–541.

Chou, S. Y., Chang, Y. H., & Shen, C. Y. (2008). A fuzzy simple additive weighting system under group decision-making for facility location selection with objective/subjective attributes. *European Journal of Operational Research, 189*(1), 132–145.

Chowdhury, R. K., & Rahman, R. (2008). Multi-criteria decision analysis in water resources management: The malnichara channel improvement. *International Journal of Environmental Science and Technology, 5*(2), 195–204.

Chung, E. S., & Lee, K. S. (2009). Identification of spatial ranking of hydrological vulnerability using multi-criteria decision making techniques: Case study of Korea. *Water Resources Management, 23*(12), 2395–2416.

Chung, E. S., & Lee, K. S. (2009). Prioritization of water management for sustainability using hydrologic simulation model and multi-criteria decision making techniques. *Journal of Environmental Management, 90*(3), 1502–1511.

Chuntian, C., & Chau, K. W. (2002). Three-person multi-objective conflict decision in reservoir flood control. *European Journal of Operational Research, 142*(3), 625–631.

Dai, F., Fu X., & Cai H. (2012). Evaluation model using the AHP of ecological environmental quality of Jiuyuangou watershed in the loess plateau, *Advanced Materials Research, 356–360*, 2571–2574

Dawes, R. M., & Corrigan, B. (1974). Linear models in decision making. *Psychological Bulletin, 81*, 95–106.

De Feo, G., & De Gisi, S. (2010). Using an innovative criteria weighting tool for stakeholders involvement to rank MSW facility sites with the AHP. *Waste Management, 30*(11), 2370–2382.

De Keyser, W., & Peeters, P. (1996). A note on the use of PROMETHEE multi-criteria methods. *European Journal of Operational Research, 89*(3), 457–461.

Delgado-GalvÃ¡n, X., et al. (2010). An analytic hierarchy process for assessing externalities in water leakage management. *Mathematical and Computer Modelling, 52*(7–8), 1194–1202.

Delle Site, P., & Filippi, F. (2009). Weighting methods in multi-attribute assessment of transport projects. *European Transport Research Review, 1*(4), 199–206.

Demirci, G., et al. (2009). In Q.M. Li et al. (Eds.), *Contractor Selection in the Housing Sector Using the Simple Multi-Attribute Rating Technique, Criocm 2009: International Symposium on Advancement of Construction Management and Real Estate* (Vols. 1-6, pp. 2672–2677). Kowloon: Hong Kong Polytechnic University

Deng, H., Yeh, C. H., & Willis, R. J. (2000). Inter-company comparison using modified TOPSIS with objective weights. *Computers and Operations Research, 27*(10), 963–973.

Diakoulaki, D., Mavrotas, G., & Papayannakis, L. (1995). Determining objective weights in multiple criteria problems: The critic method. *Computers and Operations Research, 22*(7), 763–770.

Ding, J. F. (2011). An integrated fuzzy TOPSIS method for ranking alternatives and its application. *Journal of Marine Science and Technology-Taiwan, 19*(4), 341–352.

Duarte, A. A., Rodrigues, G. M., & Ramos R. A. (2009). Effect of criteria weighting methods on the ranking of water suppliers' performance. In N. Mastorakis (Ed), Energy, environment, ecosystems, development and landscape architecture (pp. 350–355).

Dyer, J. S. (1990). Remarks on the analytic hierarchy process. *Management Science, 36*, 249–258.

Eckenrode, R. T. (1965). Weighting multiple criteria. *Management Science, Series A, Sciences, 12* (3), 180–192.

Edwards, W., & Barron, F. H. (1994). SMARTS and SMARTER: improved simple methods for multiattribute utility measurement. *Organizational Behavior and Human Decision Processes, 60*(3), 306–325.

Edwards, W. M. (1977). How to use multiattribute utility measurement for social decision making. *IEEE Transactions on Systems, Man and Cybernetics, 7*, 326–340.

Eskandari, M., Homaee, M., & Mahmodi, S. (2012). An integrated multi criteria approach for landfill siting in a conflicting environmental, economical and socio-cultural area. *Waste Management, 32*(8), 1528–1538.

Fattahi, P., & Fayyaz, S. (2010). A Compromise Programming Model to Integrated Urban Water Management. *Water Resources Management, 24*(6), 1211–1227.

Ferretti, V. (2011). A Multicriteria Spatial Decision Support System Development for Siting a Landfill in the Province of Torino (Italy). *Journal of Multi-Criteria Decision Analysis, 18*(5–6), 231–252.

Figueira, J., & Roy, B. (2002). Determining the weights of criteria in the ELECTRE type methods with a revised Simos' procedure. *European Journal of Operational Research, 139*, 317–326.

Fontana, M., Morais, D., & de Almeida, A. (2011). In R. Takahashi et al. (Eds.), *A MCDM model for urban water conservation strategies adapting Simos procedure for evaluating alternatives intra-criteria evolutionary multi-criterion optimization* (pp. 564–578). Berlin, Heidelberg: Springer.

Gallego-Ayala, J., & Juízo, D. (2011). Strategic implementation of integrated water resources management in Mozambique: An A'WOT analysis. *Physics and Chemistry of the Earth, Parts A/B/C, 36*(14–15), 1103–1111.

Gallego-Ayala, J. (2012). Selecting irrigation water pricing alternatives using a multi-methodological approach. *Mathematical and Computer Modelling, 55*(3–4), 861–883.

García, F., Guijarro, F., & Moya, I. (2010). A goal programming approach to estimating performance weights for ranking firms. *Computers and Operations Research, 37*(9), 1597–1609.

García, F., Guijarro, F., & Moya, I. (2010). Ranking Spanish savings banks: A multi-criteria approach. *Mathematical and Computer Modelling, 52*(7–8), 1058–1065.

Gómez-Limón, J. A., & Riesgo, L. (2004). Irrigation water pricing: differential impacts on irrigated farms. *Agricultural Economics, 31*(1), 47–66.

Gómez-Limón, J. A., Arriaza, M., & Riesgo, L. (2003). An MCDM analysis of agricultural risk aversion. *European Journal of Operational Research, 151*(3), 569–585.

Gonçalves, J. M., & Pereira, L. S. (2009). Decision support system for surface irrigation design. *Journal of Irrigation and Drainage Engineering, 135*, 343–356.

Gu, L., et al. (2012). Fuzzy Hierarchy comprehensive evaluation of water quality in different Lake Regions of Taihu Lake. In H.X.Q.J.Z.D. Li (Ed), Progress in Environmental Science and Engineering (pp. 844–850).

Guitouni, A., & Martel, J. M. (1998). Tentative guidelines to help choosing an appropriate MCDA method. *European Journal of Operational Research, 109*, 501–521.

Guo, Y. & Yu, Z. (2011). *The comprehensive evaluation of railway freight enterprises' performance based on the balanced scorecard card and AHP*. International Conference on Management and Service Science (MASS), 12–14 August, Wuhan, China, pp. 1–4

Hajkowicz, S. A., & Higgins, A. J. (2008). A comparison of multiple criteria analysis techniques for water resource management. *European Journal of Operational Research, 184*(1), 255–265.

Hajkowicz, S. A., & Collins, K. (2007). A review of multiple criteria analysis for water resource planning and management. *Water Resources Management, 21*(9), 1553–1566.

Hajkowicz, S. A., McDonald, G. T., & Smith, P. N. (2000). An evaluation of multiple objective decision support weighting techniques in natural resource management. *Journal of Environmental Planning and Management, 43*(4), 505–518.

Hämäläinen, R. P., & Alaja, S. (2008). The threat of weighting biases in environmental decision analysis. *Ecological Economics, 68*(1–2), 556–569.

Harmancioglu, N. B., Fedra, K., & Barbaros, F. (2008). Analysis for sustainability in management of water scarce basins: the case of the Gediz River Basin in Turkey. *Desalination, 226*(1–3), 175–182.

Hayashi, K. (2000). Multicriteria analysis for agricultural resource management: A critical survey and future perspectives. *European Journal of Operational Research, 122*(2), 486–500.

Hobbs, B., & Meier, P. (2000). *Energy decisions and the environment: A guide to the use of multicriteria methods*. Dordrecht: Kluwer Academic Publishers.

Hobbs, B. F. (1980). A Comparison of weighting methods in power plant siting. *Decision Sciences, 11*, 725–737.

Hokkanen, J., & Salminen, P. (1997). ELECTRE III and IV decision aids in an environmental problem. *Journal of Multi-criteria Decision Analysis, 6*, 215–226.

Huang, S. (2011). Designing utility-based recommender systems for e-commerce: Evaluation of preference-elicitation methods. *Electronic Commerce Research and Applications, 10*(4), 398–407.

Huber, G. P. (1974). Multi-attribute utility models: A review of field and field-like studies. *Management Science, 20*(10), 1393–1402.

Hwang, C. L., & Yoon, K. (1981). *Multiple attributes decision making methods and applications*. Berlin: Springer, Heidelberg.

Jahan, A., et al. (2012). A framework for weighting of criteria in ranking stage of material selection process. *International Journal of Advanced Manufacturing Technology, 58*(1–4), 411–420.

Janssen, R. (2001). On the use of multi-criteria analysis in environmental impact assessment in the Netherlands. *Journal of Multi-Criteria Decision Analysis, 10*, 101–109.

Jia, Z., Fan, Z. & Chen, S. (2011), Fuzzy Comprehensive Evaluation of Peak Load Unit Based on Entropy Weight. In R. Chen (Ed), *Intelligent Computing and Information Science* (pp. 737–743). Berlin, Heidelberg: Springer.

Jiang, R., et al. (2011). Research and application on flood disaster comprehensive evaluation system based on GRA. International Conference on Computer Science and Network Technology (ICCSNT), (Vol. 3, p. 2099–2102). 24–26 December, Harbin, China.

Joubert, A., Stewart, T. J., & Eberhard, R. (2003). Evaluation of water supply augmentation and water demand management options for the City of Cape Town. *Journal of Multi-Criteria Decision Analysis, 12*(1), 17–25.

Joubert, A. R., Leiman, A., de Klerk, H. M., Katau, S., & Aggenbach, J. C. (1997). Fynbos vegetation and the supply of water: a comparison of multi-criteria decision analysis and cost benefit analysis. *Ecological Economics, 22*, 123–140.

Kabak, O., & Uelengin, F. (2010). A demand forecasting methodology for fuzzy environments. *Journal of Universal Computer Science, 16*(1), 121–139.

Kamal, A. (2012). Subjective and objective weighting methods in multiple criteria decision making used in water resources, Project Report for Master of Engineering (Civil-Hydraulics and Hydrology), Faculty of Civil Engineering, Universiti, Teknologi Malaysia.

Karamouz, M., Zahraie, B., & Kerachian, R. (2003). Development of a master plan for water pollution control using MCDM techniques: a case study. *Water International, 28*(4), 478–490.

Karimi, A. R., et al. (2011). Selection of wastewater treatment process based on the analytical hierarchy process and fuzzy analytical hierarchy process methods. *International Journal of Environmental Science and Technology, 8*(2), 267–280.

Kaya, T., & Kahraman, C. (2011). Fuzzy multiple criteria forestry decision making based on an integrated VIKOR and AHP approach. *Expert Systems with Applications, 38*(6), 7326–7333.

Keeney, R. L., & Raiffa, H. (1976). *Decisions with multiple objectives: preferences and value trade offs*. New York: Wiley

Kenyon, W. (2007). Evaluating flood risk management options in Scotland: A participant-led multi-criteria approach. *Ecological Economics, 64*(1), 70–81.

Kim, J., et al. (2010). Multi-criteria decision analysis to assess options for managing contaminated sediments: application to Southern BusanHarbor, South Korea. *Integrated Environmental Assessment and Management, 6*(1), 61–71.

Kodikara, P. N., Perera, B. J. C., & Kularathna, M. D. U. P. (2010). Stakeholder preference elicitation and modelling in multi-criteria decision analysis—A case study on urban water supply. *European Journal of Operational Research, 206*(1), 209–220.

Lamy, F., et al. (2002). Development and evaluation of multi-objective decision making methods for watershed management planning. *JAWRA Journal of the American Water Resources Association, 38*(2), 517–529.

Latinopoulos, D. (2009). Multicriteria decision-making for efficient water and land resources allocation in irrigated agriculture. *Environment, Development and Sustainability, 11*(2), 329–343.

Lee, G. L., & Chan, E. W. (2008). The analytic hierarchy process (AHP) approach for assessment of urban renewal proposals. *Social Indicators Research, 89*(1), 155–168.

Levy, J. K., et al. (2007). Multi-criteria decision support systems for flood hazard mitigation and emergency response in urban watersheds1. *JAWRA Journal of the American Water Resources Association, 43*(2), 346–358.

Liu, C., et al. (2006). Catchment-wide wetland assessment and prioritization using the multi-criteria decision-making method TOPSIS. *Environmental Management, 38*(2), 316–326.

Liu, L., et al. (2010). Using fuzzy theory and information entropy for water quality assessment in three Gorges region, China. *Expert Systems with Applications, 37*(3), 2517–2521.

Macharis, C. (2005). The importance of stakeholder analysis in freight transport. *European Transport, 25–26*, 114–126.

Macharis, C., Verbeke, A., & De Brucker, K. (2004). The strategic evaluation of new technologies through multi-criteria analysis: the advisors case. In E. Bekiaris & Y. Nakanishi (Eds.), *Economic impacts of intelligent transportation systems*. Amsterdam: Innovations and case studies, Elsevier.

Machiwal, D., Jha, M. K., & Mal, B. C. (2011). Assessment of groundwater potential in a semi-arid region of india using remote sensing, GIS and MCDM techniques. *Water Resources Management, 25*(5), 1359–1386.

Malczewski, J. (1999). *GIS and Multicrieria decision analysis*. New York, NY: Wiley.

Manoliadis, O., Pantouvakis, J., & Tsolas, I. (2007). Compromise programming model in site selection for construction temporary facilities. *Operational Research, 7*(3), 381–400.

Meyar-Naimi, H., & Vaez-Zadeh, S. (2012). Sustainable development based energy policy making frameworks, a critical review. *Energy Policy, 43*, 351–361.

Miller, G. A. (1956). The magical number seven, plus or minus two: Some limits on our capacity for processing information. *The Psychological Review, 63*, 81–97.

Montazar, A., & Gaffari, A. (2012). An AHP model for crop planning within irrigation command areas. *Irrigation and Drainage, 61*(2), 168–177.

Montazar, A., & Zadbagher, E. (2010). An analytical hierarchy model for assessing global water productivity of irrigation networks in Iran. *Water Resources Management, 24*(11), 2817–2832.

Montazar, A., & Behbahani, S. M. (2007). Development of an optimised irrigation system selection model using analytical hierarchy process. *Biosystems Engineering, 98*(2), 155–165.

Morais, D. C., & Almeida, A. T. (2010). Water network rehabilitation: A group decision-making approach. *Water SA, 36*(4), 487–493.

Morais, D. C., & Almeida, A. T. (2006). Water supply system decision making using multi-criteria analysis. *Water SA, 32*(2), 229–235.

Morais, D. C., & de Almeida, A. T. (2012). Group decision making on water resources based on analysis of individual rankings. *Omega, 40*(1), 42–52.

Morais, D. C., & de Almeida, A. T. (2007). Group decision-making for leakage management strategy of water network, resources. *Conservation and Recycling, 52*(2), 441–459.

Munda, G., Nijkamp, P., & Rietveld, P. (1994). Qualitative multicriteria evaluation for environmental management. *Ecological Economics, 10*, 97–112.

Munda, G. (1995). *Multicriteria evaluation in a fuzzy environment*. Heidleberg: Physica-Verlag.

Mustajoki, J., Hämäläinen, R. P., & Marttunen, M. (2004). Participatory multi-criteria decision analysis with Web-HIPRE: A case of lake regulation policy. *Environmental Modelling and Software, 19*(6), 537–547.

Mutikanga, H., Sharma, S., & Vairavamoorthy, K. (2011). Multi-criteria decision analysis: A strategic planning tool for water loss management. *Water Resources Management, 25*(14), 3947–3969.

Mutikanga, H.E. (2012), Water loss management: tools and methods for developing countries. Ph. D. Thesis. *UNESCO-IHE Institute for Water Education*. Delft, The Netherlands.

Myllyviita, T., et al. (2012). Assessing environmental impacts of biomass production chains—application of life cycle assessment (LCA) and multi-criteria decision analysis (MCDA). *Journal of Cleaner Production, 29*(30), 238–245.

Nieto-Morote, A., Ruz-Vila, F., & Cánovas-Rodríguez, F. J. (2011). Selection of a trigeneration system using a fuzzy AHP multi-criteria decision-making approach. *International Journal of Energy Research, 35*(9), 781–794.

Nijkamp, P., Rietveld, P., & Voogd, H. (1990). *Multicriteria evaluation in physical planning*. Amsterdam: North-Holland.

Jaber, J. O., & Mohsen M. S. (2001). Evaluation of non-conventional water resources supply in Jordan. *Desalination, 136*(1–3), 83–92.

Olson, D. L. (2004). Comparison of weights in TOPSIS models. *Mathematical and Computer Modelling, 40*(7–8), 721–727.

Opricovic, S. (2009). A compromise solution in water resources planning. *Water Resources Management, 23*(8), 1549–1561.

Özcan, T., Elebi, N., & Esnaf, A. (2011). Comparative analysis of multi-criteria decision making methodologies and implementation of a warehouse location selection problem. *Expert Systems with Applications, 38*(8), 9773–9779.

Panagopoulos, G. P., et al. (2012). Mapping urban water demands using multi-criteria analysis and GIS. *Water Resources Management, 26*(5), 1347–1363.

Peacock, S. J., et al. (2007). Priority setting in health care using multi-attribute utility theory and programme budgeting and marginal analysis (PBMA). *Social Science and Medicine, 64*(4), 897–910.

Peng, H., & Zhou, H. (2011). A fuzzy-dependent chance multi-objective programming for water resources planning of a coastal city under fuzzy environment. *Water and Environment Journal, 25*(1), 40–54.

Peniwati, K., & Brenner, W. (2008). Multi-decisions rating model: establishing rescue policies for regional drinking water companies (PDAMs) in Indonesia. *European Journal of Operational Research, 186*(3), 1127–1136.

Pietersen, K. (2006). Multiple criteria decision analysis (MCDA): A tool to support sustainable management of groundwater resources in South Africa. *Water SA, 32*(2), 119–128.

Pöyhönen, M., & Hämäläinen, R. P. (2001). On the convergence of multiattribute weighting methods. *European Journal of Operational Research, 129*(3), 569–585.

Prato, T., & Herath, G. (2007). Multiple-criteria decision analysis for integrated catchment management. *Ecological Economics, 63*(2–3), 627–632.

Prato, T. (1999). Risk-based multiattribute decision making in property and watershed management. *Natural Resource Modeling, 12*(3), 307–334.

Qureshi, M., & Harrison, S. (2003). Application of the analytic hierarchy process to riparian revegetation policy options. *Small-Scale Forestry, 2*(3), 441–458.

Qureshi, M. E., & Harrison, S. R. (2001). A decision support process to compare Riparian revegetation options in Scheu Creek catchment in North Queensland. *Journal of Environmental Management, 62*(1), 101 112.

Ramjeawon, T., & Beerachee, B. (2008). Site selection of sanitary landfills on the small island of Mauritius using the analytical hierarchy process multi-criteria method. *Waste Management and Research, 26*(5), 439–447.

Rao, R. V., & Patel, B. K. (2010). A subjective and objective integrated multiple attribute decision making method for material selection. *Materials and Design, 31*(10), 4738–4747.

Rao, R. V., Patel, B. K., & Parnichkun, M. (2011). Industrial robot selection using a novel decision making method considering objective and subjective preferences. *Robotics and Autonomous Systems, 59*(6), 367–375.

Rouyendegh, B. D., & Erkan, T. E. (2012). An application of the fuzzy electre method for academic staff selection. *Human Factors and Ergonomics in Manufacturing & Service Industries, 23*(2), 107–115.

Roy, B. (1996). *Multicriteria methodology for decision aiding*. Dordecht: Kluwer Academic Publishers.

Roy, B., & Bouyssou, D. (1993). Aide multicriteere aa la deecision: Meethodes et cas, Economica, Collection Gestion, Paris.

Roy, B., & Mousseau, V. (1996). A theoretical framework for analysing the notion of relative importance of criteria. *Journal of Multi-Criteria Decision Analysis, 5*, 145–149.

RPA, Risk and Policy Analysts Ltd. (2004). Evaluating a multi-criteria analysis (MCA) methodology for application to flood management and coastal defence appraisals Case studies report Joint Defra/EA Flood and Coastal Erosion Risk Management R&D Programme.

Saaty, T. L. (1994). *The analytic hierarchy process*. New York: Wiley.

Saaty, T. L. (1987). Concepts, theory, and techniques rank generation, preservation, and reversal in the analytic hierarchy decision process. *Decision Sciences, 18*, 157–177.

Saaty, T. L. (1994). Highlights and critical points in the theory and application of the analytic hierarchy process. *European Journal of Operational Research, 74*, 426–447.

Sargaonkar, A. P., Rathi, B., & Baile, A. (2011). Identifying potential sites for artificial groundwater recharge in sub-watershed of River Kanhan, India. *Environmenta Earth Sciences, 62*(5), 1099–1108.

Schaarlig, A. (1996). Pratiquer Electre et Prometheee—Un compleement aa deecider sur plusieurs criteeres. Presses Polytechniques et Universitaires Romandes, Collection Diriger L'Entreprise, Lausanne.

Schramm, F., & Morais, D. C. (2012). Decision support model for selecting and evaluating suppliers in the construction industry. *Pesquisa Operacional, 32*(3), 643–662.

Schuwirth, N., Reichert, P., & Lienert, J. (2012). Methodological aspects of multi-criteria decision analysis for policy support: A case study on pharmaceutical removal from hospital wastewater. *European Journal of Operational Research, 220*(2), 472–483.

Seabrooke, W., Goodman, S., & Jaffry, S. (1997). Environmental valuation: using public surveys to estimate the value of natural resources'. In *Proceedings of Cutting Edge 1997, RICS Foundation.* London: RICS.

Shanian, A., et al. (2008). A new application of ELECTRE III and revised Simos' procedure for group material selection under weighting uncertainty. *Knowledge-Based Systems, 21*(7), 709–720.

Sharifi, M.A, Boerboom, L., & Shamsudin K. (2004). Evaluating rail network options using multicriteria decision analysis: Case study Klang Valley, Malaysia. In: *Application of planning and decision support systems* (pp. 21–60). Malaysia: International Islamic University.

Shemshadi, A., et al. (2011). A fuzzy VIKOR method for supplier selection based on entropy measure for objective weighting. *Expert Systems with Applications, 38*(10), 12160–12167.

Silva, V. B. S., Morais, D. C., & Almeida, A. T. (2010). A multi-criteria group decision model to support watershed committees in Brazil. *Water Resources Management, 24*(14), 4075–4091.

Simonovic, S. P., & Verma, R. (2008). A new methodology for water resources multi-criteria decision making under uncertainty. *Physics and Chemistry of the Earth, 33*(5), 322–329.

Simos, J. (1990a). L'eevaluation environnementale: Un processus cognitif neegociee. Theese de doctorat, DGF-EPFL, Lausanne.

Simos, J. (1990b). Evaluer l'impact sur l'environnement: Une approche originale par l'analyse multicriteere et la neegociation. Presses Polytechniques et Universitaires Romandes, Lausanne.

Singh, V. P. (2000). The entropy theory as a tool for modelling and decision-making in environmental and water resources. *Water SA, 26*(1), 1–10.

Sorvari, J., & Seppälä, J. (2010). A decision support tool to prioritize risk management options for contaminated sites. *Science of the Total Environment, 408*(8), 1786–1799.

Srdjevic, B. (2007). Linking analytic hierarchy process and social choice methods to group decision-making in water management. *Decision Support Systems, 42*(4), 2261–2273.

Srdjevic, B., Medeiros, Y. D. P., & Faria, A. S. (2004). An objective multi-criteria evaluation of water management scenarios. *Water Resources Management, 18*(1), 35–54.

Stewart, T. J. (1992). A critical survey on the status of multiple criteria decision making theory and practice. *OMEGA International Journal of Management Science, 20*(5/6), 569–586.

Strager, M. P. (2002). Integrating Criteria Preferences and Spatial Data to Prioritize Lands For Preservation in the Cacapon River Watershed, West Virginia, Final Report to the Canaan Valley Institute, Natural Resource Analysis Center, West Virginia University.

Straton, A., et al. (2011). Exploring and evaluating scenarios for a River Catchment in Northern Australia using scenario development, multi-criteria analysis and a deliberative process as a tool for water planning. *Water Resources Management, 25*(1), 141–164.

Supriyasilp, T., Pongput, K., & Boonyasirikul, T. (2009). Hydropower development priority using MCDM method. *Energy Policy, 37*(5), 1866–1875.

Taylor, J. G., & Ryder, S. D. (2003). Use of the Delphi method in resolving complex water resources issues. *Journal of the American Water Resources Association, 39*(1), 183–189.

Tervonen, T., Figueira, J. R., Lahdelma, R., Almeida Dias, J., & Salminen, P. (2009). A stochastic method for robustness analysis in sorting problems. *European Journal of Operational Research, 192*(1), 236–242.

Toosi, S., & Samani, J. (2012). Evaluating water transfer projects using analytic network process (ANP). *Water Resources Management, 26*(7), 1999–2014.

Valiris, G., Chytas, P., & Glykas, M. (2005). Making decisions using the balanced scorecard and the simple multi-attribute rating technique. *Performance Measurement and Metrics, 6*(3), 159–171.

van Pelt, M. J. F. (1993). Ecologically sustainable development and project appraisal in developing countries. *Ecological Economics, 7*, 19–42.

Vincke, P. (1992). *Multicriteria decision-aid.* Bruxelles, Belgium: Wiley.

Von Winterfeldt, D., & Edwards, W. (1986). *Decision analysis and behavioural research.* Cambridge: Cambridge University Press.

Voogd, H. (1983). *Multicriteria evaluation for urban and regional planning*. Londres: Pion.

Wainer, H. (1976). Estimating coefficients in linear models: It don't make no never mind. *Psychological Bulletin, 83*(2), 213–217.

Wang, J. J., et al. (2008). Using the fuzzy multi-criteria model to select the optimal cool storage system for air conditioning. *Energy and Buildings, 40*(11), 2059–2066.

Wang, J. J., Jing, Y. Y., Zhang, C. F., & Zhao, J. H. (2009), Review on multi-criteria decision aid in sustainable energy decision-making. *Renewable and Sustainable Energy Reviews, 13*, 2263–2278.

Wang, L., & Men, B. (2007). Application of TOPSIS method based on variation coefficient weight on water resource classification. In F. R. Zhang (Ed), *Proceedings of the 2007 International Conference on Agriculture Engineering*, pp. 24–27.

Wang, S. Q., Wee, Y. P., & Ofori, G. (2002). DSSDSS: A decision support system for dewatering systems selection. *Building and Environment, 37*(6), 625–645.

Wang, Y., et al. (2012). Risk assessment of floor water inrush in coal mines based on secondary fuzzy comprehensive evaluation. *International Journal of Rock Mechanics and Mining Sciences, 52*, 50–55.

Wang, Y. M., & Luo, Y. (2010). Integration of correlations with standard deviations for determining attribute weights in multiple attribute decision making. *Mathematical and Computer Modelling, 51*(1–2), 1–12.

Wang, Z., & Zhan, W. (2012). Dynamic engineering multi-criteria decision making model optimized by entropy weight for evaluating bid. *Systems Engineering Procedia, 5*, 49–54.

Whitfield, M. (1999). Prioritizing issues and concerns. A users guide to the interactive paired comparison program. Prepared for the USDA-Natural Resources Conservation Service.

Wu, H. Y., et al. (2012). Evaluation for the ecological quality status of coastal waters in East China Sea using fuzzy integrated assessment method. *Marine Pollution Bulletin, 64*(3), 546–555.

Xevi, E., & Khan, S. (2005). A multi-objective optimisation approach to water management. *Journal of Environmental Management, 77*(4), 269–277.

Xiang, X. (2008). In D. Li, (Ed), *Optimized Selection of Wetland Water Quality Monitoring Points Based on Information Entropy and Fuzzy Similarity Computer And Computing Technologies In Agriculture* (Vol. II, pp. 983–991). Boston: Springer.

Xue, D., Zhao, Q., & Guo, X. (2008). *TOPSIS method for evaluation customer service satisfaction to fast food industry.* IEEE International Conference on Service Operations and Logistics, and Informatics, (Vol. 1, pp. 920–925) 12–15 October, Beijing, China.

Yang, J. L., et al. (2008). Vendor selection by integrated fuzzy MCDM techniques with independent and interdependent relationships. *Information Sciences, 178*(21), 4166–4183.

Yang, Y., Li, B., & Yao, R. (2010). A method of identifying and weighting indicators of energy efficiency assessment in Chinese residential buildings. *Energy Policy, 38*(12), 7687–7697.

Yanhui, L., et al. (2012). Study on water resource vulnerability evaluation of Hani Terrace Core Area in Yuanyang, Yunnan. *Procedia Earth and Planetary Science, 5*, 268–274.

Yeh, C. H., & Chang, Y. H. (2009). Modeling subjective evaluation for fuzzy group multicriteria decision making. *European Journal of Operational Research, 194*(2), 464–473.

Yoe, C. (2002). Trade-Off Analysis Planning and Procedures Guidebook, Report No. IWR 02-R-2, April 2002, U.S. Army Corps of Engineers Institute for Water Resources, Alexandria, VA

Yu, V. F., & Hu, K. J. (2010). An integrated fuzzy multi-criteria approach for the performance evaluation of multiple manufacturing plants. *Computers and Industrial Engineering, 58*(2), 269–277.

Yurdusev, M., & O'Connell, P. (2005). Environmentally-sensitive water resources planning. *Water Resources Management, 19*(4), 375–397.

Zardari, N. H. (2008). An improved multicriterion analysis approach to avoid subjectivity in irrigation water allocation decisions, Ph.D. Thesis. *School of Civil and Environmental Engineering*. Sydney, Australia: University of New South Wales.

Zardari, N. H., Cordery, I., & Sharma, A. (2010). An objective multiattribute analysis approach for allocation of scarce irrigation water resources. *JAWRA Journal of the American Water Resources Association, 46*(2), 412–428.

Zarghami, M., Abrishamchi, A., & Ardakanian, R. (2008). Multi-criteria decision making for integrated urban water management. *Water Resources Management, 22*(8), 1017–1029.

Zhang, D., Yu, P. L., & Wang, P. Z. (1992). State-dependent weights in multicriteria value functions. *Journal of Optimization Theory and Applications, 74*, 1–21.

Zhang, H., et al. (2011). The evaluation of tourism destination competitiveness by TOPSIS and information entropy—A case in the Yangtze River Delta of China. *Tourism Management, 32*(2), 443–451.

Zhang, H. (2009). The analysis of the reasonable structure of water conservancy investment of capital construction in China by AHP method. *Water Resources Management, 23*(1), 1–18.

Zou, Z. H., Yun, Y., & Sun, J. N. (2006). Entropy method for determination of weight of evaluating indicators in fuzzy synthetic evaluation for water quality assessment. *Journal of Environmental Sciences, 18*(5), 1020–1023.

MIX
Papier aus verantwortungsvollen Quellen
Paper from responsible sources
FSC® C105338

If you have any concerns about our products,
you can contact us on
ProductSafety@springernature.com

In case Publisher is established outside the EU,
the EU authorized representative is:
Springer Nature Customer Service Center GmbH
Europaplatz 3, 69115 Heidelberg, Germany

Printed by Libri Plureos GmbH
in Hamburg, Germany